Two Manuals Combined into One Book

Marine Corps Martial Arts

And

The Army Weight Control Program for all Personnel

Department of Defense

Pentagon
Publishing

DEPARTMENT OF THE NAVY
Headquarters United States Marine Corps
Washington, D.C. 20380-1775

18 February 1999

FOREWORD

1. PURPOSE

Today's Marines operate within a continuum of force where conflict may change from low intensity to high intensity over a matter of hours. Marines are also engaged in many military operations other than war, such as peacekeeping missions or noncombatant evacuation operations, where deadly force may not be authorized. During noncombative engagements, Marines must determine if a situation warrants applying deadly force. Sometimes Marines must decide in a matter of seconds because their lives or the lives of others depend on their actions. To make the right decision, Marines must understand both the lethal and nonlethal close combat techniques needed to handle the situation responsibly without escalating the violence unnecessarily. Marine Corps Reference Publication (MCRP) 3-02B, *Close Combat*, provides the tactics, techniques, and procedures of Marine Corps close combat. It also provides the doctrinal basis for the Marine Corps Close Combat Training Program (MCCCTP).

2. SCOPE

This publication guides individual Marines, unit leaders, and close combat instructors in the proper tactics, techniques, and procedures for close combat. MCRP 3-02B is not intended to replace supervision by appropriate unit leaders and close combat instruction by qualified instructors. Its role is to ensure standardization and execution of tactics, techniques, and procedures throughout the Marine Corps.

3. SUPERSESSION

MCRP 3-02B supersedes Fleet Marine Force Manual (FMFM) 0-7, *Close Combat*, dated 9 July 1993. There are significant differences between the two publications. MCRP 3-02B should be reviewed in its entirety.

4. WARNING

Techniques described in this publication can cause serious injury or death. Practical application in the training of these techniques will be conducted in strict accordance with approved Entry Level Close Combat, Close Combat Instructor (CCI), and Close Combat Instructor Trainer (CCIT) lesson plans. Where serious danger exists, the reader is alerted by the following:

WARNING

5. CERTIFICATION

Reviewed and approved this date.

BY DIRECTION OF THE COMMANDANT OF THE MARINE CORPS

J. E. RHODES
Lieutenant General, U.S. Marine Corps
Commanding General
Marine Corps Combat Development Command

DISTRIBUTION: 144 000066 00

OVERVIEW OF CLOSE COMBAT

1. Purpose of Close Combat

Close combat is the physical confrontation between two or more opponents. It involves armed and unarmed and lethal and nonlethal fighting techniques that range from enforced compliance to deadly force. The purpose of close combat is to execute armed and unarmed techniques to produce both lethal and nonlethal results. Unarmed techniques include hand-to-hand combat and defense against hand-held weapons. Armed techniques include techniques applied with a rifle, bayonet, knife, baton, or any weapon of opportunity.

2. Continuum of Force

Marines will find themselves in both combative and noncombative situations. The threat level in these situations can rise and fall several times based on the actions of both Marines and the people involved. The escalation of force stops when one person complies with the demands imposed by another person. This range of actions is known as a continuum of force. Continuum of force is the concept that there is a wide range of possible actions, ranging from voice commands to application of deadly force, that may be used to gain and maintain control of a potentially dangerous situation (MCO 5500.6_, *Arming of Security and Law Enforcement [LE] Personnel and the Use of Force*). Continuum of force consists of five levels that correspond to the behavior of the people involved and the actions Marines use to handle the situation (see the table below). Close combat techniques are executed in levels three, four, and five.

Level One: Compliant (Cooperative)

The subject complies with verbal commands. Close combat techniques do not apply.

Level Two: Resistant (Passive)

The subject resists verbal commands but complies immediately to any contact controls. Close combat techniques do not apply.

Continuum of Force		
Level	**Description**	**Actions**
1	Compliant (Cooperative)	Verbal commands
2	Resistant (Passive)	Contact controls
3	Resistant (Active)	Compliance techniques
4	Assaultive (Bodily Harm)	Defensive tactics
5	Assaultive (Serious Bodily Harm/ Death)	Deadly force
Note: Shading indicates levels in which Marines use close combat techniques.		

Level Three: Resistant (Active)

The subject initially demonstrates physical resistance. Marines use compliance techniques to control the situation. Level three incorporates close combat techniques to physically force a subject to comply. Techniques include—

- Come-along holds.
- Soft-handed stunning blows.
- Pain compliance through joint manipulation and the use of pressure points.

Level Four: Assaultive (Bodily Harm)

The subject may physically attack Marines, but he does not use a weapon. Marines use defensive tactics to neutralize the threat. Defensive tactics include the following close combat techniques:

- Blocks.
- Strikes.
- Kicks.
- Enhanced pain compliance procedures.
- Nightstick blocks and blows.

Level Five: Assaultive (Serious Bodily Harm/Death)

The subject usually has a weapon and will either kill or seriously injure someone if he is not stopped immediately and brought under control. Typically, to control the subject, Marines apply deadly force through the use of a firearm, but they may also use armed and unarmed close combat techniques.

3. Marine Corps Tactical Concepts

Close combat techniques support the following key Marine Corps tactical concepts. The concepts are not standalone ideas but are to be combined to achieve an effect that is greater than their separate sum.

Achieving a Decision

Achieving a decision is important in close combat. An indecisive fight wastes energy and possibly Marines' lives. Whether the intent is to control an opponent through restraint or defend themselves in war, Marines must have a clear purpose before engaging in close combat and act decisively once engaged.

Gaining an Advantage

A basic principle of martial arts is to use the opponent's strength and momentum against him to gain more leverage than one's own muscles alone can generate, thereby gaining an advantage. In close combat, Marines must exploit every advantage over an opponent to ensure a successful outcome. This can include employing various weapons and close combat techniques that will present a dilemma to an opponent. Achieving surprise can also greatly increase leverage. Marines try to achieve surprise through deception, stealth, and ambiguity.

Speed

Marines use speed to gain the initiative and advantage over the enemy. In close combat, the speed and violence of the attack against an opponent provides Marines with a distinct advantage. Marines must know and understand the basics of close combat so they can act instinctively with speed to execute close combat techniques.

Adapting

Close combat can be characterized by friction, uncertainty, disorder, and rapid change. Each situation is a unique combination of shifting factors that cannot be controlled with precision or certainty. For example, a crowd control mission may call for Marines to employ various techniques ranging from nonlethal restraint to more forceful applications. Marines who adapt quickly will have a significant advantage.

Exploiting Success

Typically, an enemy will not normally surrender simply because he was placed at a disadvantage. Marines cannot be satisfied with gaining an advantage in a close combat situation. They must exploit any advantage aggressively and ruthlessly until an opportunity arises to completely dominate the opponent. Marines must exploit success by using every advantage that can be gained.

Close Combat

Table of Contents

Page

Page

CHAPTER 1

FUNDAMENTALS OF CLOSE COMBAT

This chapter describes all techniques for a right-handed person. However, all techniques can be executed from either side.

The Marine is depicted in camouflage utilities. The opponent is depicted without camouflage.

The fundamentals of close combat include ranges, weapons of the body, target areas of the body, and pressure points of the body. These fundamentals form the basis for all close combat techniques. They provide Marines with a common framework regardless of the type of confrontation or the techniques used. If Marines apply these fundamentals properly in a close combat situation, they may save their lives or the lives of fellow Marines.

1. Ranges of Close Combat

Close combat engagements occur within three ranges: long range, midrange, and close range. During any engagement, these ranges may blur together or may rapidly transition from one to another until either the opponent is defeated or the conflict is resolved.

Long Range

During long range engagements, combatants engage each other with rifles, bayonets, sticks, or entrenching tools. See figure below.

Midrange

During midrange engagements, combatants engage each other with knives, punches, or kicks.

Close Range

During close range engagements, combatants grab each other. Close range engagements also involve elbow strikes, knee strikes, and grappling.

2. Weapons of the Body

Hands and Arms

The hands, forearms, and elbows are the arm's individual weapons. The hands consist of several areas that can be used as weapons: fists, edges of hands, palms, and fingers.

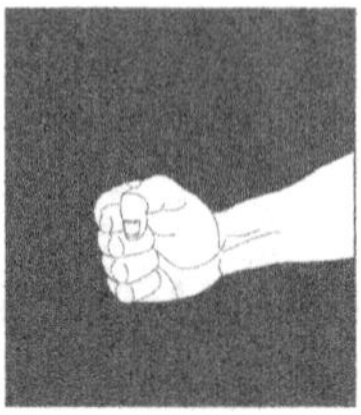

Fists. To minimize injury to the fists, Marines use their fists as weapons to target soft tissue areas such as the throat. The fists' striking surfaces are the first two knuckles of the hands or the meaty portions of the hands below the little fingers.

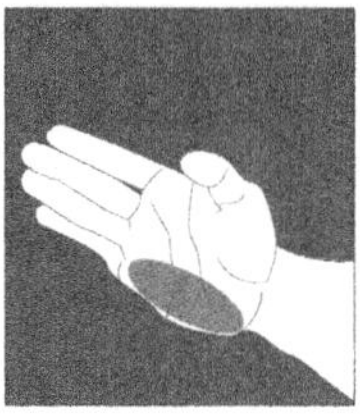

Edge of Hand. Marines use the edge of the hand (knife edge) as a weapon. Marines use the edge of the hand to strike soft tissue areas.

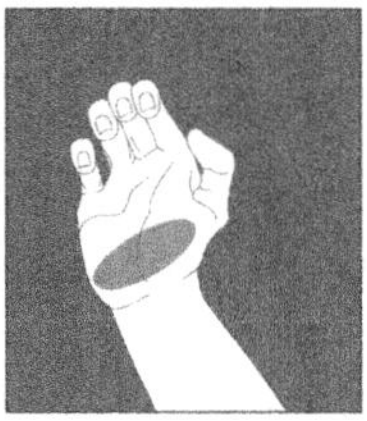

Palms. Because of the palm's padding, Marines use the heels of the palms to strike, parry, and/or block.

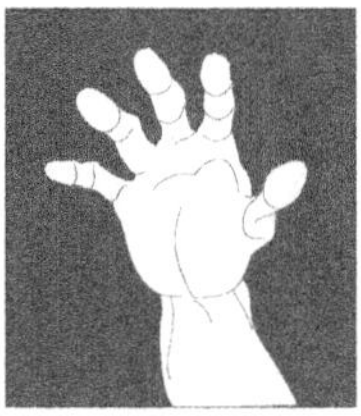

Fingers. Marines use the fingers to gouge, rip, and tear soft tissue areas (e.g., eyes, throat, groin).

Forearms. Marines use the forearms as a defensive tool to deflect or block attacks. Forearms can also be used as striking weapons to damage or break an opponent's joints and limbs. Marines sustain less self-injury when strikes are conducted with the forearms than when strikes are conducted with fists and fingers.

Elbows. Marines use the elbows as striking weapons. Because of the short distance needed to generate power, elbows are excellent weapons for striking during the close range of close combat.

Legs

The legs are more powerful than any other weapon of the body, and they are less prone to injury when striking. The feet are protected by boots and are the preferred choice for striking.

Feet. Marines use the balls of the feet, the insteps, and the toes to kick an opponent. Marines use the cutting edge of the heels and the heels to stomp on an opponent. Marines must be wearing boots when striking with the toes.

Knees. Like elbows, knees are excellent weapons in the close range of close combat. Knee strikes are most effective while fighting close to an opponent where kicks are impractical. The opponent's groin area is an ideal target for the knee strike if he is standing upright. Knee strikes can deliver a devastating secondary attack to an opponent's face following an initial attack that caused him to bend at the waist.

3. Target Areas of the Body

During close combat, Marines strive to attack the accessible target areas of an opponent's body. The readily accessible areas will vary with each situation and throughout the engagement. The target areas are divided into five major groups: head, neck, torso, groin, and extremities. The figure below illustrates target areas of the body.

Head

The vulnerable regions of the head are the eyes, temple, nose, ears, and jaw. Massive damage to the head kills an opponent.

Eyes. The eyes are excellent targets because they are soft tissue areas that are not protected by bone or muscle. Attacks to this area may cause the opponent to protect the area with his hands, allowing Marines to execute a secondary attack to other

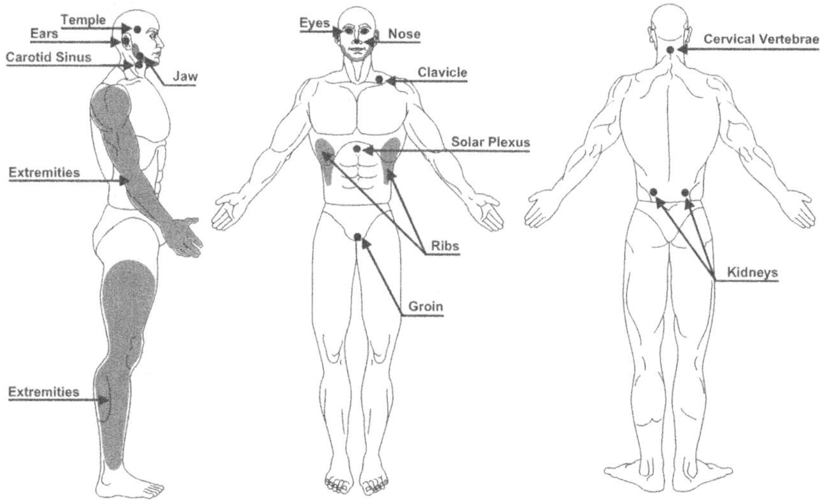

target areas while the opponent uses his hands to protect his eyes.

Temple. The temple is one of the most fragile areas of the head. Powerful strikes to the opponent's temple cause permanent damage and death.

Nose. The nose is very sensitive and easily broken. An attack to this area causes involuntary watering and closing of the opponent's eyes, rendering him vulnerable to secondary attacks. However, through training, individuals can condition themselves to withstand attacks to the nose. Therefore, any attack to the nose must be powerfully delivered.

Ears. Attacks to the ears may cause the eardrum to rupture. But this may not stop or even distract an opponent unless Marines powerfully deliver the strike.

Jaw. The jaw region, when struck forcefully, renders the opponent unconscious. Strikes to the jaw cause painful injuries to the teeth and surrounding tissues (e.g., lips, tongue), but the risk of self-injury is great unless Marines deliver strikes with a hard object such as a helmet, rifle butt, or boot heel.

Neck

The front of the neck, or throat area, is a soft tissue area that is not covered by natural protection. Damage to this region causes the opponent's trachea to swell, closing his airway, which can lead to death.

Carotid Sinus. The carotid sinus is located on both sides of the neck just below the jaw. Strikes to the carotid sinus restrict blood flow to the brain, causing loss of consciousness or death.

Cervical Vertebrae. The cervical vertebrae on the back of the neck, from the base of the skull to the top of the shoulders, contains the spinal cord, which is the nervous system's link to the brain. The weight of the head and the lack of large muscle mass allow damage to the cervical vertebrae and spinal cord. Excessive damage to this area causes pain, paralysis, or death.

Torso

Clavicle. The opponent's clavicle (or collar bone) can be easily fractured, causing immobilization of the arm.

Solar Plexus. Attacks to the opponent's solar plexus or center of the chest can knock the breath out of him and immobilize him.

Ribs. Damage to the opponent's ribs immobilizes him. It may also cause internal trauma.

Kidneys. Powerful attacks to the opponent's kidneys cause immobilization, permanent damage, or death.

Groin

The groin area is another soft tissue area not covered by natural protection. Any damage to this area causes the opponent to involuntarily protect his injured area, usually with his hands or legs. In male opponents, the scrotum is the main target since even a near miss causes severe pain, contraction of the lower abdominal muscles, deterioration of his stance, and possible internal trauma.

Extremities

Rarely will an attack to the opponent's extremities (arms and legs) cause death, but they are important target areas in close combat. Damage to an opponent's joints causes immobilization.

4. Pressure Points of the Body

There are nerves in the human body that, when pressure is applied or when they are struck, allow Marines to control a subject through pain compliance. Marines use pressure points to control an opponent when deadly force is not authorized. They also use pressure points to soften or distract an opponent so a lethal or nonlethal technique can be employed. The figure on page 1-5 illustrates the body's pressure points. Marines execute attacks to pressure points by—

- Rapidly kicking or striking pressure points.
- Slowly applying steady pressure to pressure points.

Infraorbital Nerve

The infraorbital nerve is just below the nose. Marines apply pressure to this nerve with an index finger to control the opponent.

Mastoid Process

The mastoid process is behind the base of the ear and beneath the edge of the jaw. Marines apply inward and upward pressure to this pressure point with the fingers to distract and control the opponent.

Jugular Notch

The jugular notch is at the base of the neck in the notch formed at the center of the clavicle. Marines apply pressure in a quick, stabbing motion with the index finger. Strikes to the jugular notch cause serious damage.

Brachial Plexus (Tie In)

The brachial plexus (tie in) is on the front of the shoulder at the joint. Strikes and pressure applied with the hand are effective on this nerve.

Radial Nerves

Radial nerves are on the inside of the forearms along the radius bones. Strikes and pressure applied with the hand to the radial nerve serve as a softening technique.

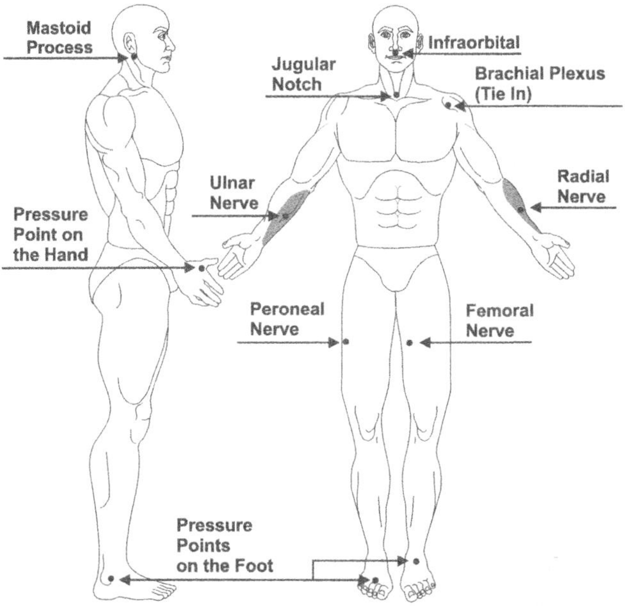

Ulnar Nerve

Ulnar nerves are on the outside of the forearms along the ulnar bones. Strikes and pressure applied with the hand to the ulnar nerve serve as a softening technique.

Pressure Point on the Hand

The hands contain a pressure point on the webbing between the thumbs and index fingers where the two bones of the fingers meet.

To force an opponent to soften or release his grip, Marines apply pressure with their index fingers to this pressure point or strike this pressure point with their fists.

Femoral Nerves

Femoral nerves are on the inside of the thighs along the femur bones. Strikes to the femoral nerve serve as a softening technique.

Peroneal Nerves

Peroneal nerves are on the outside of the thighs along the femur bones. Strikes to the peroneal nerve serve as a softening technique.

Pressure Points on the Feet

There are pressure points on the feet that, when pressure is applied or when they are struck, serve to soften or distract the opponent. Marines apply pressure with the toe, edge, or heel of their boots to the following points:

- The notch below the ball of the ankle.
- The top center of the foot, above the toes.
- The top of the foot where the leg and foot meet.

5. Basic Warrior Stance

Achieving a solid stance is critical to stability and movement throughout any close combat confrontation. The basic warrior stance provides the foundation for stability and movement that is needed to execute close combat techniques. To execute the basic warrior stance, Marines put their feet apart, hands up, elbows in, and chin down.

Feet Apart

Place feet shoulder-width apart.

Keep the head forward and eyes on the opponent, take a half step forward with the left foot, and pivot on the heels so the hips and shoulders are at approximately a 45-degree angle to the right.

Distribute body weight evenly on both legs. Bend the knees slightly.

Hands Up

Curl the fingers naturally into the palm of the hand. Position the thumb across the index and middle fingers. Do not clench the fists. Clenching the fists increases muscular tension in the forearms and decreases speed and reaction time.

Bring the hands up to the face at chin level, with the palms facing each other. Hold the fists up high enough to protect the head, but not so high that they block the field of vision. Ensure continuous eye contact with the opponent.

Elbows In

Tuck the elbows in close to protect the body.

Chin Down

Tuck the chin down to take advantage of the natural protection provided by the shoulders.

6. Angles of Approach and Movement

Marines use movement to control a confrontation and to retain a tactical advantage. Movement increases power and maximizes momentum. By moving around the opponent, Marines gain access to different target areas of the opponent's body. Movement allows Marines to use different weapons of the body and different close combat techniques to attack specific target areas.

Angles of Approach

Marines move anywhere within a 360-degree circle around the opponent to gain a tactical advantage. This circle provides access to different target areas of the opponent's body.

When facing an opponent, Marines move in a 45-degree angle to either side of the opponent. Moving at a 45-degree angle avoids an opponent's strike and puts Marines in the best position to attack the opponent. Marines should avoid being directly in front of an opponent during a confrontation. If a Marine is directly in front of an opponent, the opponent can rely on his forward momentum and linear power to seize the tactical advantage.

Movement

Marines must know how to move in all directions while maintaining the basic warrior stance. During any movement, the legs or feet should not be crossed. Once a movement is completed, the basic warrior stance should be resumed. Maintaining the basic warrior stance protects Marines and puts them in the proper position to launch an attack against an opponent.

Note: Before body movement begins, Marines turn their heads quickly to the new direction. The faster the head turns, the faster the body moves, and the quicker Marines attain visual contact with the opponent.

Forward to the Left. To move forward to the left from the basic warrior stance, Marines—

- Move the left foot forward at a 45-degree angle from the body (approximately 12 to 15 inches), keeping the toe pointed toward the opponent.
- Bring the right foot behind the left foot as soon as the left foot is in place. This returns Marines to the basic warrior stance.

Forward to the Right. To move forward to the right from the basic warrior stance, Marines—

- Move the right foot forward at a 45-degree angle from the body (approximately 12 to 15 inches).
- Bring the left foot, toe pointing toward the opponent, in front of the right foot as soon as the right foot is in place. This returns Marines to the basic warrior stance.

Backward to the Left. To move backward to the left from the basic warrior stance, Marines execute the forward movement in reverse. Marines—

- Move the left foot backward at a 45-degree angle from the body (approximately 12 to 15 inches), keeping the toe pointed toward the opponent.
- Bring the right foot behind the left foot as soon as the left foot is in place. This returns Marines to the basic warrior stance.

Backward to the Right. To move backward to the right from the basic warrior stance, Marines execute the forward movement in reverse. Marines—

- Move the right foot backward at a 45-degree angle from the body (approximately 12 to 15 inches).
- Bring the left foot, toe pointing toward the opponent, in front of the right foot as soon as the right foot is in place. This returns Marines to the basic warrior stance.

7. Balance and Off-Balancing

Balance

In any close combat situation, Marines must strive to maintain balance. The last place to be in a close combat situation is on the ground. Marines must maintain a strong base and a low center of balance, their feet must be a shoulder-width apart, and they must stay on their toes to enable quick movement.

Off-balancing

Marines use off-balancing techniques to control an opponent. These techniques are used to throw an opponent to the ground while Marines remain standing, or they are used to put Marines in a position for an offensive attack.

Off-balancing techniques use the opponent's momentum to move or throw him. For example, if the opponent is charging a Marine, the Marine pulls the opponent to drive him to the ground. Likewise, if the opponent is pulling a Marine, the Marine pushes the opponent to drive him to the ground.

Off-balancing techniques also rely on the power generated by the opponent. For example, during combat a Marine may be tired or outnumbered. Depending on the generated energy and momentum of the opponent, the Marine employs off-balancing techniques with very little effort and still provides effective results.

Because off-balancing techniques rely on the momentum and power generated by the opponent, these techniques are particularly effective for Marines who may be outsized by their opponent or lack their opponent's strength.

Angles of Off-balancing

There are eight angles or directions in which an opponent can be off-balanced: forward, rear, right, left, forward right, forward left, rear right, and rear left.

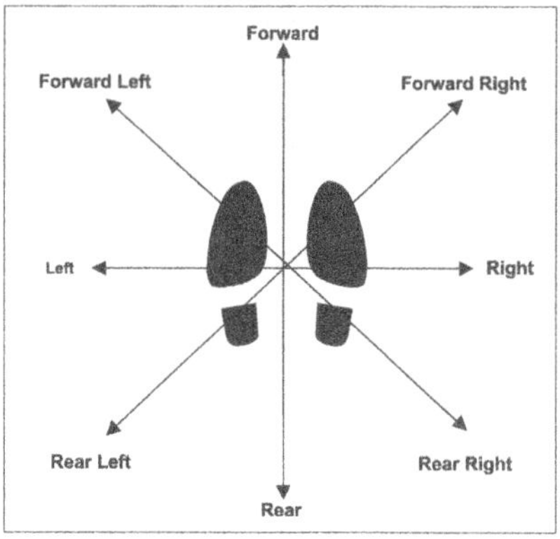

Note: The angles correspond to the Marine's perspective, not the opponent's. Forward, rear, right, and left are straight angles. Forward right, forward left, rear right, and rear left are considered quadrants that are at a 45-degree angle in either direction to the front or rear.

Off-balancing Techniques

Marines off-balance an opponent by pushing, pulling, or bumping the opponent with their hands, arms, or bodies.

To pull, Marines grab an opponent with their hands and drive him forcefully to one of the rear quadrants or to the right or left.

To push, Marines grab the opponent with their hands and drive him forcefully into one of the front quadrants or to the right or left.

Marines execute bumping in the same manner as pushing, but use their shoulders, hips, and legs instead of their hands to off-balance the opponent.

8. Falls

Marines may lose their balance or be thrown to the ground during encounters with an opponent. Marines use falling techniques to absorb the impact of a fall and to quickly return to their feet following an opponent's attack.

Whether falling or being thrown by an opponent, Marines strive to reduce the force of the impact, to prevent serious personal injury, and to increase the chances of survival. Falling techniques use the body's large muscles (back, thighs, buttocks) to protect vital organs and bones from injury and immobilization.

Front Fall

Marines execute a front fall to break a fall on the front. To execute the front fall, Marines—

Bend the elbows and place the palms facing out in a position to spread and absorb the impact of the fall.

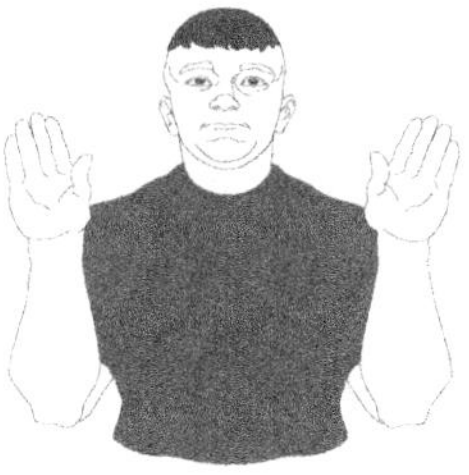

Fall forward, breaking the fall with the forearms and palms. The forearms and hands, down to the fingertips, should strike the ground simultaneously.

Offer resistance with the forearms and hands to keep the head raised off the ground.

Side Fall

Marines execute a side fall to break a fall on the side. To execute the side fall, Marines—

Bring the right arm across the body so the hand is next to the left shoulder with the palm facing inboard.

Fall to the side, breaking the fall with the right arm by slapping the ground and making contact from the shoulder or forearm down to the hand. At the same time, tuck the chin and keep the head raised off the ground. The chin should be tucked to the chest at all times to prevent whiplash.

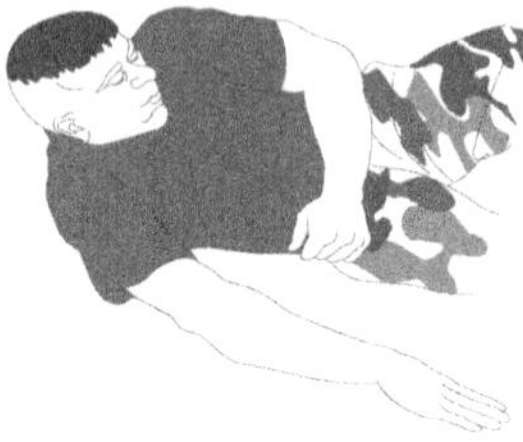

Stretch the right leg out to make contact with the ground and to distribute and absorb the impact. Bend the left leg, allowing the foot to make contact with the ground.

Back Fall

Marines execute a back fall to break the fall when being thrown or falling backward. To execute the back fall, Marines—

Cross the hands in front of the chest and tuck the chin.

Fall backward and slap the ground with the forearms and hands to absorb the impact of the fall and keep the head off the ground.

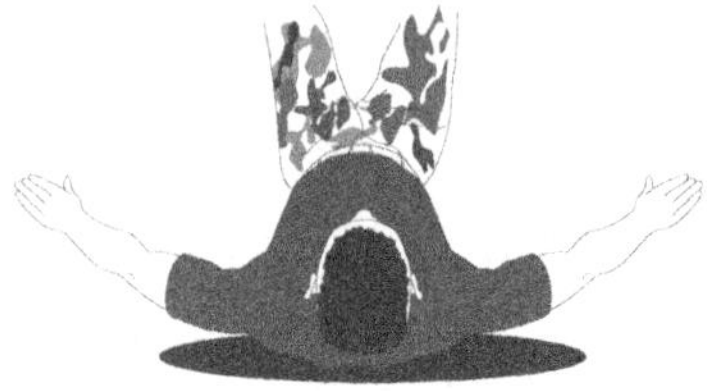

Forward Shoulder Roll

Marines use the forward shoulder roll to break a fall from an opponent's attack and to use the momentum of the fall to get back on their feet quickly. Ideally, Marines execute the forward shoulder roll to a standing position so they can continue fighting. To execute the forward shoulder roll, Marines—

Contact the ground with the back of the right forearm and upper arm. Tuck the chin into the chest.

Roll onto the right shoulder, rolling diagonally across the back to land on the left hip.

Slap the ground with the left arm, absorbing the impact from the shoulder to the hand, palm down.

Keep the left leg straight to absorb as much of the impact as possible. The right leg is bent and the foot hits flat on the ground.

Bend the left leg upon impact to push off with the left knee and leg to a squatting and then a standing position. Forward momentum should carry the Marine to a standing position.

CHAPTER 2

LETHAL AND NONLETHAL WEAPONS TECHNIQUES

This chapter describes all techniques for a right-handed person. However, all techniques can be executed from either side.

In drawings, the Marine is depicted in woodland camouflage utilities; the opponent is depicted without camouflage. In photographs, the Marine is depicted in woodland camouflage utilities; the opponent is depicted in desert camouflage utilities.

1. Bayonet Techniques

WARNING

During training, Marines must have bayonets sheathed. Marines use bayonet dummies to practice bayonet techniques. When practicing offensive and defensive bayonet techniques student-on-student, Marines use pugil sticks.

All Marines armed with a rifle carry a bayonet. The bayonet is an effective weapon if Marines are properly trained in offensive and defensive bayonet techniques. An offensive attack, such as a thrust, is a devastating attack that can quickly end a fight. Defensive techniques, such as the block and parry, can deter the opponent's attack and allow Marines to regain the initiative. Through proper training, Marines develop the courage and confidence required to effectively use a bayonet to protect themselves and destroy the enemy. In situations where friendly and enemy troops are closely mingled and rifle fire and grenades are impractical, the bayonet becomes the weapon of choice.

Holding the Rifle

To execute bayonet techniques, Marines hold the rifle in a modified basic warrior stance. All movement begins and ends with the basic warrior stance. To hold the rifle, Marines—

Use an overhanded grasp to grab the small of the rifle's stock. Use an underhanded grasp to grab the handguards of the rifle.

Lock the buttstock of the rifle against the hip with the right forearm.

Orient the blade end of the rifle toward the opponent.

Offensive Bayonet Techniques

Straight Thrust. Marines use the straight thrust to disable or kill an opponent. It is the most deadly offensive technique because it causes the most trauma to an opponent. Target areas are the opponent's throat, groin, or face. The opponent's chest and stomach are also excellent target areas if not protected by body armor or combat equipment. To execute the straight thrust, Marines—

Lift the left leg and lunge forward off the ball of the right foot while thrusting the blade end of the weapon forward, directly toward the opponent.

Retract the weapon and return to the basic warrior stance.

Slash. Marines use the slash to kill an opponent or to create an opening in his defense. The target area is the opponent's neck. To execute the slash, Marines—

Extend the left hand back toward the left shoulder.

Thrust the left hand forward and swing it to the right, bring the right hand back toward the hip, and turn the cutting edge of the blade toward the opponent's neck. The movement is a slashing motion so the blade cuts across the opponent's neck.

Horizontal Buttstroke. Marines use the horizontal buttstroke to weaken an opponent's defenses, to cause serious injury, or to set him up for a killing blow. Target areas are the opponent's head, neck, and legs. To execute the horizontal buttstroke, Marines—

Step forward with the right foot and drive the right hand forward. Rotate the hips and shoulders into the strike. Move the left hand back toward the left shoulder.

Strike the opponent with the butt of the weapon.

Vertical Buttstroke. Marines use the vertical buttstroke to weaken an opponent's defenses, to cause serious injury, or to set him up for a killing blow. Target areas are the opponent's groin and face. To execute the vertical buttstroke, Marines—

Step forward with the right foot and drive the right hand straight up.

Pull the left hand back over the left shoulder.

Strike the opponent with the butt of the weapon.

Smash. Marines use the smash as a follow-up technique to the horizontal or vertical buttstroke, particularly if they missed the target. The target area is the opponent's head. To execute the smash following a buttstroke, Marines—

Step forward with the right foot and place the blade end of the weapon over the left shoulder and elevate the right elbow above the shoulders.

Drive the arms straight forward, striking the opponent with the butt of the weapon.

Defensive Bayonet Techniques

Parry. Marines use a parry as a defensive technique to redirect or deflect an attack. A parry is a slight redirection of a linear attack by an opponent; e.g., a straight thrust or a smash. To execute the parry, Marines—

Use the bayonet end of the rifle to redirect the barrel or bayonet of the opponent's weapon.

Lock the weapon against the hip with the right forearm.

Rotate to the right or left, moving the bayonet end of the rifle to parry the opponent's attack. Rotation is generated from the hips.

Redirect or guide the opponent's weapon away from the body by exerting pressure against the opponent's weapon.

High Block. Marines execute a high block against a vertical attack coming from high to low. To execute the high block, Marines—

Thrust the arms up forcefully at approximately a 45-degree angle from the body. The weapon should be over the top of the head and parallel to the ground. The elbows are bent, but there is enough muscular tension in the arms to absorb the impact and deter the attack.

Low Block. Marines execute the low block

against a vertical attack coming from low to high. To execute the low block, Marines—

Thrust the arms down forcefully at approximately a 45-degree angle from the body. The weapon

should be at or below the waist and parallel to the ground. The elbows are bent, but there is enough muscular tension in the arms to absorb the impact and deter the attack.

Left and Right Block. Marines execute a left or right block against a horizontal buttstroke or a slash. To execute the left or right block, Marines—

Thrust the arms forcefully to the right or left, holding the rifle vertically in the direction of the attack. The elbows are bent, but there is enough muscular tension in the arms to absorb the impact and deter the attack.

Counter Action Following the Block. After deflecting an opponent's attack with a block, Marines counter with a slash or a horizontal buttstroke to regain the initiative. However, the objective in any bayonet fight is to thrust forward with the blade end of the weapon to immediately end the fight.

Group Strategy

On occasion, Marines may engage an opponent as a member of a group or numerous opponents by one's self or as a member of a group. By combining bayonet fighting movements and simple strategies, Marines can effectively overcome their opponent or opponents.

Offensive Strategy: Two Against One. If two bayonet fighters engage one opponent, the fighters advance together.

Fighter 1 engages the opponent while fighter 2 swiftly and aggressively attacks the opponent's exposed flank and destroys the opponent.

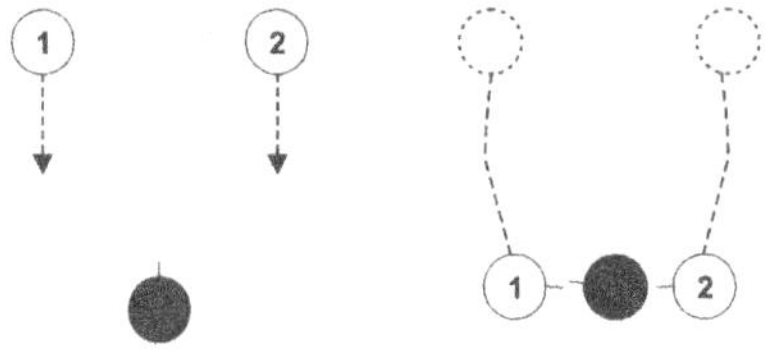

Offensive Strategy: Three Against Two. If three bayonet fighters engage two opponents, the fighters advance together keeping their opponents to the inside.

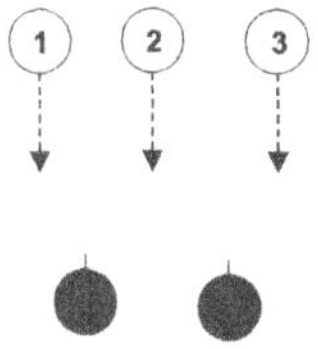

Fighters 1 and 3 engage opponents. Fighter 2 attacks the opponent's exposed flank engaged by fighter 1 and destroys the opponent.

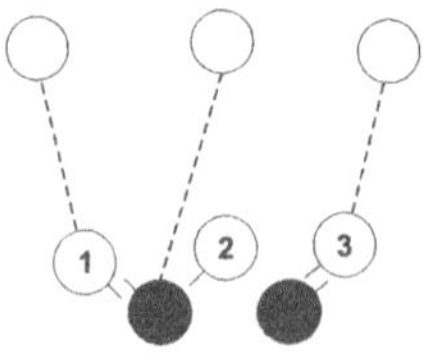

Fighters 1 and 2 turn and attack the exposed flank of the opponent engaged by fighter 3 and destroy the opponent.

Defensive Strategy: One Against Two. If a fighter is attacked by two opponents, the fighter immediately positions himself at the flank of the nearest opponent and keeps that opponent between himself and the other opponent.

Using the first opponent's body as a shield against the second opponent, the fighter destroys the first opponent quickly before the second opponent moves to assist.

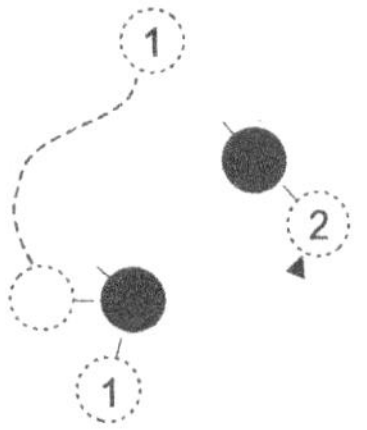

Then, the fighter engages and destroys the second opponent.

Defensive Strategy: Two Against Three. If two fighters are attacked by three opponents, the fighters immediately move to the opponent's flanks.

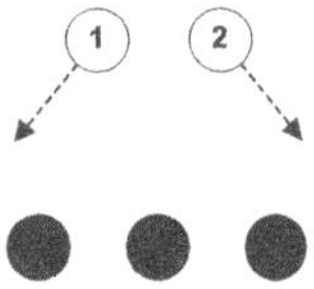

Fighters 1 and 2 quickly attack and destroy their opponents before the third opponent closes in.

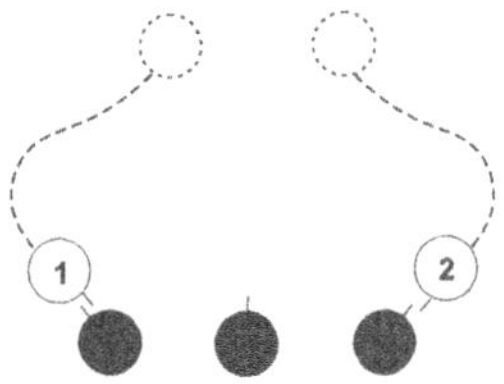

Fighter 1 engages the third opponent while fighter 2 attacks the opponent's exposed flank and destroys the opponent.

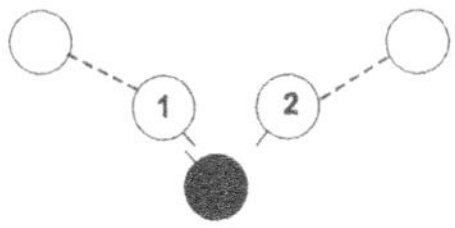

2. Nonlethal Rifle and Shotgun Retention Techniques

Most Marines are armed with the M16A2 service rifle. Marines are taught to keep their weapon with them at all times. Marines must be constantly alert to their surroundings and the people moving in and around their environment. Marines may be confronted by an individual who tries to take their weapons. If this happens, Marines should not struggle with the individual. To retain positive control of their weapons, Marines must understand and apply weapons retention techniques, otherwise known as armed manipulation. The following techniques can be used with either the rifle or the shotgun.

Blocking Technique

To execute a blocking technique, Marines—

Stand in a defensive posture.

Use the weapon to block the opponent by thrusting it out firmly, with the elbows bent. Do not try to hit the opponent with the rifle. The rifle is used as a barrier.

Technique if Opponent Grabs Weapon Underhanded

Usually an opponent will try to grab the weapon or block it as an instinctive action. If the opponent uses an underhand grab to seize the handguards of the weapon, Marines—

Trap the opponent's closest finger(s) above the knuckle with the thumb so he cannot release his grip.

Apply bone pressure on the opponent's finger to initiate pain compliance.

Rotate the barrel of the weapon up or down quickly while maintaining pressure on the opponent's hand. At the same time, quickly pivot to off-balance the opponent.

Technique if Opponent Grabs Weapon Overhanded

If the opponent uses an overhand grab to seize the weapon overhanded, Marines—

Trap the opponent's finger to hold his hand in place.

Rotate the barrel to place it across the opponent's forearm and apply downward pressure. This action is similar to an armbar.

Technique if the Opponent Grabs the Muzzle

If the opponent grabs the muzzle of the weapon, Marines—

Rotate the muzzle quickly in a circle motion. Slash downward with the muzzle to release the opponent's grip.

Butt Strikes

Strikes with the butt of the weapon control or ward off an attacker. During any of the retention techniques, Marines use the heel or cutting edge of the weapon to deliver butt strikes to the inside or the outside of the opponent's thigh. The inside butt strike targets the femoral nerve as illustrated in the photo to the left.

The outside butt strike targets the peroneal nerve as illustrated in the photo to the left. Strikes can be made to the outside or inside of the thighs. If a strike to one side of the thigh misses, Marines follow back through with the butt of the weapon on the other side of the thigh.

Off-balancing Techniques

Marines apply off-balancing techniques to throw an opponent to the ground and retain possession of the weapon.

If the opponent grabs the weapon and pushes, Marines should not push on the weapon. They should—

Move with the momentum and movement of the

opponent by pivoting in the direction of the movement by stepping back.

Throw the opponent to the ground with a quick jerking movement by lowering the muzzle and swinging the butt of the weapon.

If the opponent grabs the weapon and pulls, Marines—

Step on the opponent's foot and push forward to off-balance him and drive him to the ground.

Sweep the opponent's feet out from under him by hooking his leg with the leg and kicking backward.

3. Nonlethal Handgun Retention Techniques

Many Marines are armed with the M9 service pistol. Marines must keep their weapons in their possession at all times. Marines must be constantly alert to their surroundings and the people moving in and around the environment. Marines may be confronted by an individual who tries to take their weapons. To retain positive control of the weapon, Marines must understand and apply handgun retention techniques.

Blocking Technique

Marines perform the following blocking techniques if an opponent attempts to grab their pistol in the holster. Marines—

Place the body between the weapon and the opponent by immediately pivoting so the weapon is away from the opponent.

Step back and away from the opponent while placing the hand on the pistol grip.

Extend the left hand and block, deflect, or strike the opponent's arm.

Armbar Technique

Marines use the armbar technique when an opponent uses his right hand to grab their pistol in the holster. To execute the armbar technique, Marines—

Trap the opponent's right hand by grasping the opponent's wrist or hand with the right hand and applying pressure against the body.

Step back with the right foot and pivot sharply to the right to be next to the opponent. Always pivot in a direction that keeps the weapon away from the opponent.

Straighten the opponent's arm to apply an armbar. The arm should be straight across the torso.

Continue pivoting to the right while pulling back on the opponent's shoulder. This action may break the opponent's arm.

Place the left hand across the opponent's face and apply pressure back and down to take the oppo-

nent to the ground. Pressure applied beneath the nose or on the trachea is equally effective.

Wristlock Technique

Marines use the wristlock technique when an opponent grabs their pistol while it is in the holster with his right hand. To execute the wristlock technique, Marines—

Grasp the opponent's wrist or hand with the right hand and apply pressure against the body.

Step back and away from the opponent and pivot to the right so the weapon is away from the opponent. Always pivot in a direction that keeps the weapon away from the opponent.

Reach over the opponent's arm with the left hand and grab his hand, applying pressure against his arm with the left forearm. Execute a wristlock.

Incorporate the second hand into the wristlock and, stepping back with the left foot, pivot to the left.

Execute a two-handed wristlock by exerting downward pressure with the thumb and rotating his hand to the left.

Continue to pivot to off-balance the opponent and drive him to the ground.

Note: If the opponent grabs the pistol with his left hand, Marines execute the wristlock with one hand and step in toward the opponent, rather than

away from the opponent.

Softening Techniques

Handgun retention techniques use softening techniques applied to pressure points. Bone pressure

and strikes with the hands (i.e., hammer fist), knees, and feet are also effective softening techniques.

Pressure Points. Marines use pressure point techniques to get the opponent to loosen his grip. Marines use their finger tips to apply pressure to the webbing between the index finger and thumb, the jugular notch, and the brachial plexus tie in. The following figure illustrates pressure applied to the brachial plexus tie in.

Bone Pressure. Bone pressure is the application of pressure on a bone against a hard object to initiate pain compliance. To apply bone pressure, Marines use their hand to trap the opponent's hand on the weapon. Marines apply a slow, steady pressure to the opponent's hand and fingers until his grip is softened or he releases his hold.

Strikes. If it is difficult to apply a retention technique, Marines employ strikes or kicks to force the opponent to loosen his grip. Strikes to the

eyes, the arms (radial nerve), or shoulder (brachial plexus tie in) soften the opponent's grip on the weapon.

Kicks and knee strikes to the peroneal nerve, the femoral nerve, or the groin are effective because the opponent is typically unprepared to counter the strike.

Stomping on the top of the opponent's foot may distract him or loosen his grip on the weapon.

4. Firearm Disarmament Techniques

Marines use firearm disarmament techniques during a close-range confrontation if they are un-

armed and the opponent has a firearm (pistol). These techniques are equally effective if Marines are armed but do not have time to withdraw and present the weapon. The goal of firearm disarmament techniques is to gain control of the situation

so Marines gain the tactical advantage. The goal is not necessarily to get control of the opponent's weapon.

Pistol to the Front

This technique is used when Marines are unarmed and the opponent has a pistol pointing at their front (e.g., head, chest). The technique is the same if the opponent sticks the pistol under the Marine's chin. To execute the counter when an opponent is pointing a pistol toward the front of a Marine, Marines—

Place the hands close to the weapon, about chest high, palms out.

Use the left hand to grab the opponent's forearm and push the opponent's hand with the pistol away to clear the body from in front of the weapon. At the same time, rotate the right shoulder back to clear the body from the weapon.

Maintain control of the opponent's arm.

Grasp the weapon with the right hand by placing the thumb underneath the pistol and the fingers over top of the pistol.

Keep the right hand wrapped tightly around the muzzle and quickly rotate the pistol in the opponent's hand so the muzzle is facing the opponent.

Grasp and pull the opponent's wrist or forearm away from the body while rotating the weapon.

Rotate the weapon toward the opponent while pulling it up and back and out of the opponent's grasp.

Pistol to the Rear

This technique is performed when Marines are unarmed and the opponent has a pistol pointing to the back of the Marine's head. To execute the counter when the opponent is pointing a pistol to the rear, Marines—

Place the hands close together about chest high, palms out.

Step back with the left foot, pivoting on the right foot so the side is against the opponent's front. This action clears the body from the weapon's line of fire. Keep the left hand up.

Pivot on the left foot to face the opponent and, at the same time, raise the left elbow and reach over the top of the opponent's arm with the left arm.

Wrap the left arm tightly around the opponent's arm above his elbow to control it.

Push on the opponent's shoulder with the right hand while pulling up with the left arm to achieve an armbar. This action releases the opponent's grip on the weapon. If necessary, the opponent can be taken to the ground with a leg sweep.

Note: To execute this technique, the weapon must be close to or touching the rear of the Marine. If the weapon is too far away from the body, this technique would be difficult to execute or it would be ineffective.

CHAPTER 3

HAND-HELD WEAPONS

This chapter describes all techniques for a right-handed person. However, all techniques can be executed from either side.

In drawings, the Marine is depicted in woodland camouflage utilities; the opponent is depicted without camouflage. In photographs, the Marine is depicted in woodland camouflage utilities; the opponent is depicted in desert camouflage utilities.

Marines must know how to defend against attacks when an opponent is either unarmed or armed with a held-held weapon. This chapter addresses the combative use of knives, specific weapons of opportunity, and sticks. However, virtually anything can be used as a hand-held weapon.

1. Fundamentals of Knife Fighting

Marines must be trained in knife fighting techniques. Marines experienced in offensive knife techniques can cause enough damage and massive trauma to stop an opponent. When engaged against each other, experienced knife fighters employ various maneuvers and techniques that are specific to knife fighting. Seldom, if ever, will Marines engage an opponent in a classical knife fight.

Note: When armed with a rifle, Marines are issued a bayonet. When armed with a pistol, Marine are issued a combat knife.

Angles of Attack

There are six angles from which an attack with a knife can be launched:

- Vertical strike coming straight down on an opponent.
- Forward diagonal strike coming in at a 45-degree angle to the opponent.
- Reverse diagonal strike coming in at a 45-degree angle to the opponent.
- Forward horizontal strike coming in parallel to the ground.
- Reverse horizontal strike coming in parallel to the ground.
- Forward thrust coming in a straight line to the opponent.

Target Areas of the Body

During any confrontation, the parts of the opponent's body that are exposed or readily accessible will vary. The goal in a knife fight is to attack the body's soft, vital target areas that are readily accessible (e.g, the face, the sides and front of the neck, the lower abdomen [or groin]).

Neck. Carotid arteries, located on either side of the neck, are good target areas because they are not covered by body armor or natural protection.

Lower Abdomen (or Groin). The lower abdomen (or groin region) is a good target area because it is not covered by body armor.

Heart. The heart, if not covered by body armor, is an excellent target which, if struck, can prove fatal in a matter of seconds or minutes.

Secondary Targets. There are secondary target areas that will cause substantial bleeding if an artery is severed. These target areas are not immediately fatal, but can become fatal if left unattended. Attacks to—

- The legs can cause a great deal of trauma and prove fatal. For example, the femoral artery located on the inside of the thigh is a large artery which, if cut, will cause extensive blood loss.
- The brachial artery, located between the biceps and triceps on the inside of the arm, can cause extensive bleeding and damage.
- The arm's radial and ulnar nerves can cause extensive bleeding and damage.

Movement

Marines can move anywhere within a 360-degree circle around the opponent. This allows accessibility to different target areas of the opponent's body. Marines should avoid being directly in front of an opponent because the opponent can rely on his forward momentum to seize the tactical advantage. If Marines face an opponent, movement is made in a 45-degree angle to either side of the opponent. This angle avoids an opponent's strike and places Marines in the best position to attack an opponent.

Wearing the Combat Knife

Marines must wear the combat knife where it is easily accessible and where it can best be retained. It is recommended the combat knife be worn on the weak side hip, blade down, sharp edge facing forward. Marines can place it behind the magazine pouch where it is easily accessible to them, but not easily grabbed by an opponent.

Grip

The grip on the knife should be natural. Marines grasp the knife's grip with the fingers wrapped around the grip naturally as it is pulled out of its sheath. This is commonly known as a hammer grip. The blade end of the knife is always facing the opponent.

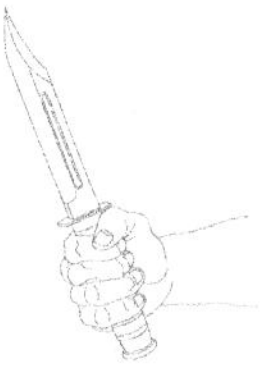

Stance

Marines use the basic warrior stance as the foundation for knife techniques. The left hand is a vertical shield that protects either the ribs or the head and neck. The right elbow is bent with the blade pointing forward toward the opponent's head. This position serves as an index point, where all techniques are initiated.

Principles of Knife Fighting

The following are key principles of knife fighting:

- Execute movements with the knife blade within a box, shoulder-width across from the neck down to the waistline. The opponent has a greater chance of blocking an attack if the blade is brought in a wide, sweeping movement to the opponent.
- Close with the opponent, coming straight to the target.
- Move with the knife in straight lines.
- Point the knife's blade tip forward and toward the opponent.
- Apply full body weight and power in each of the knife techniques. Full body weight should be put into the attack in the direction of the blade's movement (slash or thrust).
- Apply constant forward pressure with the body and blade to keep the opponent off-balanced.

2. Knife Fighting Techniques

Slashing Techniques

Marines use slashing techniques to close with an enemy. Slashing techniques distract the opponent or damage the opponent so Marines can close in. Typically, Marines target the opponent's limbs, but any portion of the body that is presented can become a target.

Vertical Slash Technique. The vertical slash follows a vertical line straight down through the target. To execute the vertical slash, Marines—

Thrust the right hand out and bring the weapon straight down on the opponent, continuing to drag the knife down through the opponent's body.

Maintain contact on the opponent's body with the blade of the knife.

Forward Slash Technique. The forward slash follows a straight line in a forehand stroke, across the target areas of either the neck (high diagonal slash) or abdominal region (low horizontal slash). To execute the forward slash, Marines—

Extend the right hand while simultaneously rotating the palm up until the knife blade makes contact with the opponent.

Snap or rotate the wrist through the slashing motion to maximize blade's contact with the opponent.

Drag the knife across the opponent's body, from right to left, in a forehand stroke. The movement ends with the forearm against the body and the knife at the left hip with its blade oriented toward the opponent.

Reverse Slash Technique. The reverse slash is a follow-up technique to a forward attack. It allows Marines both a secondary attack and the ability to resume the basic warrior stance. The reverse slash follows a straight line in a backhand stroke, across the target areas of either the neck (high diagonal slash) or abdominal region (low horizontal slash). To execute the reverse slash, Marines—

Extend the right hand while simultaneously rotating the palm down until the knife blade makes contact with the opponent.

Snap or rotate the wrist through the slashing motion to maximize the blade's contact with the opponent.

Drag the knife across the opponent's body, from left to right, in a backhand stroke. Maintain contact on the opponent's body with the blade of the knife.

Thrusting Techniques

The primary objective of knife fighting is to insert the blade into an opponent to cause extensive damage and trauma. This is done with a thrusting technique. Thrusting techniques are more effective than slashing techniques because of the damage they can inflict. However, Marines use slashing techniques to close with the enemy so that they are closer to the opponent, which allows them to use the thrusting technique.

Vertical Thrust. The thrusting motion follows a vertical line straight up through the target (low into the abdomen region or high into the neck). To execute the vertical thrust, Marines—

Thrust the right hand toward the target, inserting the knife blade straight into the opponent.

Pull the knife out of the opponent.

Forward Thrust. The forward thrust follows a straight line straight into the opponent's neck (high thrust) or abdominal region (low thrust). To execute the forward thrust, Marines—

Thrust the right hand, palm down, toward the target, inserting the knife blade straight into the opponent.

Rotate the palm up once the knife is inserted to twist the blade.

Drop the right elbow and bring the knife to the opposite side of the opponent's body from where it was inserted. At the same time, rotate the hips and shoulders downward to bring body weight to bear on the attack.

Reverse Thrust. The reverse thrust is a follow-up technique to a forward attack. It allows Marines both a secondary attack and the ability to resume the basic warrior stance. The reverse thrust follows a horizontal line straight into the opponent's neck (high thrust) or abdominal region (low thrust). To execute the reverse thrust, Marines—

Bend the right arm, crossing the arm to the left side of the body.

Thrust the right hand, palm up, toward the target, inserting the knife blade straight into the opponent.

Rotate the palm down to twist the blade once the knife is inserted.

Bring the knife to the opposite side of the opponent's body from where it was inserted.

3. Weapons of Opportunity

During an unarmed close combat situation, Marines use their bodies as weapons, but they should be ready and able to use anything around them as a weapon. For example, Marines could throw sand or liquid in an opponent's eyes to temporarily impair his vision or smash the opponent's head with a rock or helmet. Marines must use whatever means are available and do whatever it takes to take control of the situation and to win, or they face the possibility of losing their lives. Some weapons of opportunity are discussed in the following subparagraphs.

Entrenching Tool

An entrenching tool (E-tool) is commonly carried by Marines. It can be an excellent weapon, especially when sharpened. Marines can use the E-tool to block, slash, and thrust at an opponent.

Tent Pole and Pins

Marines can use tent poles and pins to block, strike, or thrust at an opponent.

Web Belt

Marines can stretch a web belt between their hands to block attacks by an opponent.

Battlefield Debris

Marines can use debris on the battlefield (e.g., sticks, glass, a sharp piece of metal) to cut, slash, or stab an opponent. They can also use other types of debris such as shovel or ax handles, boards, metal pipes, or broken rifles to strike an opponent or apply a choke.

Helmet

A helmet can be used to strike an opponent on an unprotected area like the head and face. Grasp the rim of the helmet and thrust the arms forward, striking the opponent with the top of the helmet.

4. Fundamentals of Combative Stick

On the battlefield, Marines must be ready and able to use anything as a weapon. They must learn and be able to use techniques that can be employed with most weapons of opportunity. Among these techniques are combative stick techniques. Combative stick techniques can be used with a stick, a club, a broken rifle, an E-tool, or even a web belt.

Angles of Attack

There are six angles from which an attack with a hand-held weapon can be launched:

- Vertical strike coming straight down on an opponent.
- Forward diagonal strike coming in at a 45-degree angle to the opponent.
- Reverse diagonal strike coming in at a 45-degree angle to the opponent.
- Forward horizontal strike coming in parallel to the ground.
- Reverse horizontal strike coming in parallel to the ground.
- Forward thrust coming in a straight line to the opponent.

Grip

Grasp the stick about 2 inches from its base.

Stance

The basic warrior stance serves as the foundation for combative stick techniques. The left hand becomes a vertical shield that protects the ribs or the head and neck. Depending on how heavy the weapon is, it should be held at a level approximately shoulder height.

Movement

Movement during combative stick techniques is the same as it is for other close combat techniques. Marines can move anywhere within a 360-degree circle around the opponent. This allows accessibility to different target areas of the opponent's body and gains a tactical advantage. Marines should avoid being directly in front of an opponent because the opponent can rely on his forward momentum to seize the tactical advantage. If Marines face an opponent, movement is made in a 45-degree angle to either side of the opponent. This angle avoids an opponent's strike and places Marines in the best position to attack an opponent.

5. Combative Stick Techniques

Strikes

Strikes are intended to inflict as much damage on an opponent as possible. Striking techniques apply to a weapon of opportunity such as a stick, a tent pole, a club, a broken rifle, an E-tool, or a pipe.

Vertical Strike. To execute the vertical strike, Marines—

Bend the right arm, extending the weapon over the back of the right shoulder.

Rotate the forearm straight down off the elbow to bring the weapon down on the opponent.

Rotate the hips and shoulders forcefully toward the opponent.

Follow through with the strike by allowing the weight of the weapon to go through the target area of the body.

Forward Strike. To execute a forward strike, Marines—

Step forward with the left foot in the direction of the strike.

Bend the right arm with the elbow extending out to the right and the weapon extended over the right shoulder.

Rotate the forearm to the right of the elbow to bring the weapon down onto the opponent. At the same time, forcefully rotate the hips and shoulders toward the opponent.

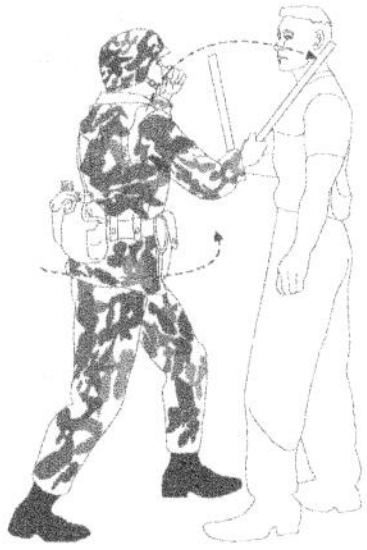

Follow through with the strike by allowing the weight of the weapon to go through the target area of the body.

Reverse Strike. The reverse strike is a follow-up technique to a forward strike. It allows Marines both a secondary attack and the ability to resume the basic warrior stance.To execute a reverse strike, Marines—

Step forward with the right foot in the direction of the strike.

Bend the right arm with the hand near the left shoulder. The weapon is extended over the left shoulder.

Rotate the forearm to the right of the elbow to bring the weapon down onto the opponent. At the same time, forcefully rotate the hips and shoulders toward the opponent.

Follow through with the strike by allowing the weight of the weapon to go through the target area of the body.

Forward Thrust. To execute the forward thrust, Marines—

Grasp the stick with the left hand, palm up, in a position where the stick can be controlled with two hands.

Lift the left leg and lunge forward off the ball of the right foot. At the same time, thrust the end of the weapon directly toward the opponent by thrusting both hands forward in a straight line.

6. Blocking Techniques

A block is meant to deter or deflect an attack by an opponent. A block sets up Marines for a follow-on attack against the opponent. Blocks are executed by deflecting, rather than hitting or following through like a strike.

Blocks Against Unarmed Attacks

To block against an unarmed attack, Marines—

Step forward at a 45-degree angle with either the right or left foot. This moves the body out of the line of attack.

Raise the left arm and block the strike with the meaty portion of the forearm.

Employ the stick by one of two techniques:

- Strike the opponent with the stick.
- Use the stick to block on two points of contact. When the stick is used to block, it serves as an incidental strike. This technique is only used if Marines have closed with the opponent and are inside his strike.

Blocks Against Armed Attacks

Block for a Vertical Strike. To execute the block for a vertical strike, Marines—

Step forward with the left foot at a 45-degree angle to the left. This moves the body out of the line of attack.

Block on two points of contact to disperse the impact of the attack:

- Block the opponent's stick by positioning the stick so it is perpendicular to the opponent's stick. If the stick is not perpendicular to the opponent's stick, the stick can slide through and make contact on the Marine.
- Block the opponent's wrist or forearm with the back of the left forearm.

Note: Use the stick to block the opponent's arm if closer to the opponent. It is the same movement as blocking with the arm, except the opponent's arm is blocked with both the stick and the arm.

Block for a Forward Strike. To execute the block for a forward strike, Marines—

Step forward with the left foot at a 45-degree angle to the left. This moves the body out of the line of attack and inside the opponent's strike.

Block on two points of contact to disperse the impact of the attack:

- Block the opponent's wrist or forearm with the meaty portion of the back of the left forearm.
- Strike the opponent's attacking biceps with the stick.

Block for a Reverse Strike. To execute the block for a reverse strike, Marines—

Step forward at a 45-degree angle with the right foot to the right. This moves the body out of the line of attack and inside the opponent's strike.

Block on two points of contact to disperse the impact of the attack:

- Block the opponent's stick by positioning the stick so it is perpendicular to the opponent's stick.

Block the opponent's forearm with the meaty portion of the left forearm.

Note: If closer to the opponent, block his triceps with the back of the left forearm and strike his forearm with the stick.

7. Unarmed Against Hand-Held Weapons

If Marines are engaged against an opponent with a knife, a stick, or some other weapon of opportunity, they must establish and maintain an offensive mindset, not a defensive mindset. Their survival depends on it. They cannot afford to think about getting cut or hurt.

Angles of Attack

Before Marines can learn to block or counter an attack with a hand-held weapon (e.g., knife, stick), they must know from what angle the opponent is attacking. There are six angles from which an opponent will typically attack with a hand-held weapon:

- Vertical strike coming in straight down on the Marine.
- Forward diagonal strike coming in at a 45-degree angle to the Marine.
- Reverse diagonal strike coming in at a 45-degree angle to the Marine.
- Forward horizontal strike coming in parallel to the ground.
- Reverse horizontal strike coming in parallel to the ground.
- Forward thrust coming in a straight line to the Marine.

Blocks

If the opponent has a hand-held weapon (e.g., knife or stick), Marines parry the opponent's hand or arm to block the attack.

Basic Block. To execute the basic block technique, Marines—

Step forward at a 45-degree angle to move out of the line of the attack. Always step in the direction of the strike.

Thrust the forearms forward, hands up, against the opponent's attacking arm. Contact is made on the opponent's arm with the backs of the fore-arms.

Block for a Vertical Strike. To execute the block against a vertical strike, Marines—

Step forward with the left foot at a 45-degree an-gle to the left to move out of the line of attack.

Thrust the forearms forward, hands up, against the outside of the opponent's attacking arm.

Block for a Forward Strike. To execute the block against a forward diagonal or forward hori-zontal strike, Marines—

Step forward with the left foot inside the oppo-nent's attacking arm.

Block the attack with both arms bent so the fore-arms make contact with the opponent's biceps and forearm.

Block for a Reverse Strike. To execute the block against a reverse diagonal or reverse hori-zontal strike, Marines—

Step forward with the right foot to the outside of the opponent's attacking arm.

Block the attack with both arms bent so the fore-arms make contact with the opponent's triceps and forearm.

Block for a Forward Thrust. To execute the block against a forward thrust, Marines—

Bend at the waist, move the hips backward, and jump backward with both feet to move away from the attack. This action is known as "hollowing out."

Hollow out and block the attack with the arms bent and hands together on top of the attacking arm.

Overlap the hands slightly so one thumb is on top of the other hand's index finger. The other thumb should be under the other hand's index finger.

8. Counters to Hand-Held Weapon Attacks

Principles of Counters

A counter is used to control the situation to regain the tactical advantage and end the fight. Regardless of the type of weapon or angle of attack, the following principles apply to countering the attack with a hand-held weapon:

- Move out of the line of attack. Movement is executed in a 45-degree angle forward to the left or right.
- Block the attack.

Note: The first two actions are taken simultaneously.

- Take control of the weapon by controlling the hand or arm that is holding the weapon. Never attempt to grab the opponent's weapon.
- Execute the appropriate follow-up to end the fight; e.g., strikes, joint manipulations, throws, or takedowns (see chap. 8). Marines should continue to attack the opponent until the fight ends.

Counter Techniques

There are two techniques that can be used to counter any armed attack: forward armbar counter and reverse armbar counter. These techniques can be used to counter a vertical attack, a forward diagonal strike, or a forward horizontal strike. With minor variations, the same techniques are used to counter reverse strikes. A third technique, the bent armbar counter, is used to counter a vertical attack.

Forward Armbar Counter. To execute the forward armbar counter to an attack coming from a forward strike, Marines—

Step forward with the left foot inside the opponent's attacking arm.

Block the attack with both arms bent so the forearms make contact with the opponent's biceps and forearm.

Slide the left arm over the opponent's forearm and wrap the arm tightly around his arm, trapping the opponent's attacking arm between the biceps and torso.

Place the right hand on the opponent's shoulder or upper arm to further control his arm and to effect an armbar.

Execute an armbar and continue to exert steady pressure against the arm.

Reverse Armbar Counter. To execute the reverse armbar counter to an attack coming from a forward strike, Marines—

Step forward with the left foot inside the opponent's attacking arm.

Block the attack with both arms bent so the forearms make contact with the opponent's biceps and forearm.

Control the opponent's arm with the left hand and pivot to the right so the back is against the opponent's side. Immediately slide the right arm over the opponent's biceps and wrap the arm tightly around his arm, trapping the opponent's attacking arm between the biceps and torso.

Grasp the opponent's wrist with the left hand and twist his thumb away from the body.

Take the opponent to the ground with an armbar takedown if he does not drop the weapon from the wristlock.

Bent Armbar Counter. This counter is particularly effective against a vertical attack. To execute the bent armbar counter, Marines—

Step forward with the left foot inside the opponent's attacking arm.

Block the attack with both arms bent so the forearms make contact with the opponent's biceps and forearm.

Grasp the opponent's forearm with the left hand. At the same time, slide the right arm underneath the opponent's triceps and grasp the opponent's forearm or wrist with the right hand.

Apply downward pressure with the hands against the opponent's forearm to off-balance the opponent.

CHAPTER 4

STRIKES

This chapter describes all techniques for a right-handed person. However, all techniques can be executed from either side.

In drawings, the Marine is depicted in woodland camouflage utilities; the opponent is depicted without camouflage. In photographs, the Marine is depicted in woodland camouflage utilities; the opponent is depicted in desert camouflage utilities.

Strikes are unarmed individual hitting techniques. Strikes use the hands, elbows, knees, feet, and, in some instances, other parts of the body as personal weapons. Marines must know how to execute strikes effectively. They must also know how to counter strikes from an opponent.

1. Principles of Punches

Muscle Relaxation

Muscle relaxation is crucial when executing punches. The natural tendency in a fight is to tense up, which results in rapid fatigue and decreased power generation. Marines who remain relaxed during a close combat situation generate greater speed, which results in greater generation of power. Relaxing the forearms generates speed and improves reaction time. At the point of impact, Marines clench the fist to cause damage to the opponent and avoid injury to the wrist and hand.

Weight Transfer

Weight transfer is necessary to generate power in a punch. Marines accomplish this by—

- Rotating their hips and shoulders into the attack.
- Moving their body mass straight forward or backward in a straight line.
- Dropping their body weight into an opponent. Body mass can be transferred into an attack from high to low or from low to high.

Rapid Retraction

When Marines deliver a punch, rapid retraction of the fist is important. Once the hand has made contact with the target, Marines quickly return to the basic warrior stance. Rapid retraction—

- Returns the hand and arm to the protection afforded by the basic warrior stance.
- Prevents the opponent from being able to grab the hand or arm.
- Permits the hand and arm to be "chambered" or "re-cocked" in preparation for delivering a subsequent punch.

Telegraphing

Telegraphing a strike occurs when body movements inform the opponent of the intention to launch an attack. Staying relaxed helps to reduce telegraphing.

Often, an untrained fighter telegraphs his intention to attack by drawing his hand back in view of his opponent, changing facial expression, tensing neck muscles, or twitching. These movements, however small, immediately indicate an attack is about to be delivered. If the opponent is a trained fighter, he may be able to evade or counter the attack. If the opponent is an untrained fighter, he may still be able to minimize the effect of an attack.

2. Punches

Punches may be thrown during any hand-to-hand confrontation. Most people resort to punching because it is a natural reaction to a threat. The purpose of a punch is to stun the opponent or to set him up for a follow-up finishing technique. However, punches should only be executed to the soft tissue areas of an opponent. A correctly delivered punch maximizes the damage to an opponent while minimizing the risk of injury to Marines.

Basic Fist

Punches are executed using the basic fist. To make the basic fist, the fingers are curled naturally into the palm of the hand and the thumb is placed across the index and middle fingers. Do not clench the fist until movement has begun. This reduces muscular tension in the forearm and increases speed and reaction time. Just before impact, Marines exert muscular tension on the hand and forearm to maximize damage to the opponent and reduce their chances of injury. Contact should be made with the knuckles of the index and middle fingers.

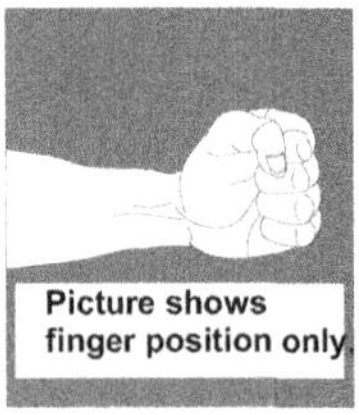

When striking with the basic fist, Marines must keep the hand straight, or in line, with the wrist to avoid injury to the wrist.

Lead Hand Punch

The lead hand punch is a snapping straight punch executed by the forward or lead hand. It is a fast punch designed to keep the opponent away and to set up an attack. A lead hand punch conceals movement and allows Marines to get close to the opponent. Lead hand punches should strike soft tissue areas, if possible. To execute the lead hand punch, Marines—

Snap the lead hand out to nearly full extension, while rotating the palm to the ground.

Contact the opponent with the first two knuckles of the fist.

Retract the hand immediately, resuming the basic warrior stance.

Rear Hand Punch

The rear hand punch is a snapping punch executed by the rear (right) hand. It is a power punch designed to inflict maximum damage on the opponent. Its power comes from pushing off the rear leg and rotating the hips and shoulders. To execute the rear hand punch, Marines—

Rotate the hips and shoulders forcefully toward the opponent and thrust the rear hand straight out, rotating the palm down, to nearly full extension.

Shift body weight to the lead foot while pushing off on the ball of the rear foot.

Contact the opponent with the first two knuckles of the fist.

Retract the hand immediately.

Uppercut

The uppercut is a powerful punch originating below the opponent's line of vision. It is executed in an upward motion traveling up the centerline of the opponent's body. It is delivered in close and usually follows a preparatory strike that leaves the target area unprotected. When delivered to the chin or jaw, the uppercut can render an opponent unconscious, cause extensive damage to the neck, or sever the tongue. To execute the uppercut, Marines—

Bend the arms, rotating the palm inboard. The distance the arms bend depends on how close the opponent is.

Rotate the hips and shoulders forcefully toward the opponent, thrusting the fist straight up toward the opponent's chin or jaw.

Contact the opponent with the first two knuckles of the fist.

Retract the hand immediately.

Hook

The hook is a powerful punch that is executed close in and is usually preceded by a preparatory strike. To execute the hook, Marines—

Thrust the right arm in a hooking motion toward the opponent, keeping the elbow bent while forcefully rotating the right shoulder and hip toward the opponent.

Contact the opponent with the first two knuckles of the fist. Continue rotating the shoulder and hip, following through with the fist to the target.

Retract the hand immediately.

3. Strikes with the Upper Body

Strikes stun the opponent or set him up for a follow-up finishing technique. The hands, forearms, and elbows are individual weapons of the arms that can be used to execute strikes including the hammer fist, chin jab, knife hand, eye gouge, and elbow strikes. These strikes provide a variety of techniques that can be used in any type of close combat situation.

Principles of Execution

There are several principles of execution that ensure a strike's effectiveness.

Generating Power. Weight transfer is necessary to generate maximum power in a punch. Marines accomplish this by—

- Rotating the hips and shoulders into the attack.
- Moving the body mass straight forward or backward in a straight line.
- Dropping body weight into an opponent. Body mass can be transferred into an attack from high to low or from low to high.

Muscular Tension. The arms are relaxed until the moment of impact. At the point of impact, Marines apply muscular tension in the hand and forearm to maximize damage to the opponent and to avoid injury to the hand. The arms are relaxed until the moment of impact.

Hit and Stick and Follow-Through. A strike should be delivered so that the weapon (e.g., hand, elbow) hits and remains on the impact site (target) and follows through the target. This technique inflicts maximum damage on the opponent.

Strikes with the arms are executed with "heavy hands"; i.e., the strike is executed by driving through with the strike to allow the weight of the hand to go through the target area of the body. Contact with the opponent should be made with the arm slightly bent, and the arm extends as it moves through the target. This technique allows Marines to deliver strikes effectively without executing full force.

Movement. Movement puts Marines in the proper position for launching an attack against an opponent as well as providing protection.

Movement is initiated from the basic warrior stance and ends by resuming the basic warrior stance. Strikes can be performed with either the left or right arm depending upon—

- The angle of attack.
- The position of the opponent.
- The opponent's available, vulnerable target areas.

Target Areas of the Body

For each strike, there are target areas of the body that, when struck, maximize damage to an opponent. Strikes use gross motor skills as opposed to fine motor skills. The target areas of the body are just that—areas. Pinpoint accuracy on a specific nerve is not needed for the strike to be effective.

Hammer Fist

Striking with the hammer fist concentrates power in a small part of the hand, which, when transferred to the target, can have a devastating effect. The striking surface of the hammer fist is the meaty portion of the hand below the little finger. To execute the hammer fist strike, Marines—

Make a fist and bend the arm at approximately a 45- to 90-degree angle. At the same time, rotate the right hip and right shoulder backward.

Thrust the fist forward onto the opponent while rotating the right hip and shoulder forward.

Rotate the wrist so the hammer fist makes contact on the opponent.

Chin Jab

The chin jab can immediately render an opponent unconscious and cause extensive damage to the neck and spine. The striking surface is the heel of the palm of the hand. To execute the chin jab, Marines—

Bend the right wrist back at a 90-degree angle with the palm facing the opponent and the fingers pointing up.

Keep the right arm bent and close to the body. Extend the hand into a concave position with the fingers slightly spread apart.

Step forward with the left foot toward the opponent, keeping the feet approximately shoulder-width apart and the knees bent. This is done to close with the opponent.

Keep the right arm bent and close to the side. Thrust the palm of the hand directly up under the opponent's chin. At the same time, rotate the right hip forward to drive the body weight into the attack to increase the power of the strike. The attack should travel up the centerline of the opponent's chest to his chin.

Knife Hand

The knife hand is one of the most versatile and devastating strikes. The striking surface is the cutting edge of the hand, which is the meaty portion of the hand below the little finger extending to the top of the wrist. The striking surface is narrow, allowing strikes on the neck between the opponent's body armor and helmet. The knife hand strike is executed from one of three angles: outside, inside, and vertical.

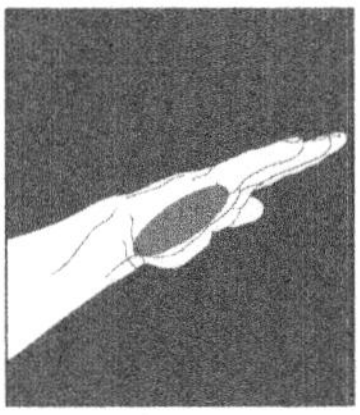

Outside Knife Hand. To execute the outside knife hand strike, Marines—

Execute a knife hand by extending and joining the fingers of the right hand and placing the thumb next to the forefinger (like saluting).

Retract the right hand. At the same time, rotate the right hip and right shoulder backward.

Thrust the knife hand forward (horizontally) onto the opponent while rotating the right hip and shoulder forward.

Inside Knife Hand. To execute the inside knife hand strike, Marines—

Execute a knife hand.

Bring the right hand over the left shoulder. At the same time, rotate the right shoulder forward and the left hip forward.

Thrust the knife hand forward (horizontally) onto the opponent while rotating the right hip and shoulder forward and the left shoulder backward.

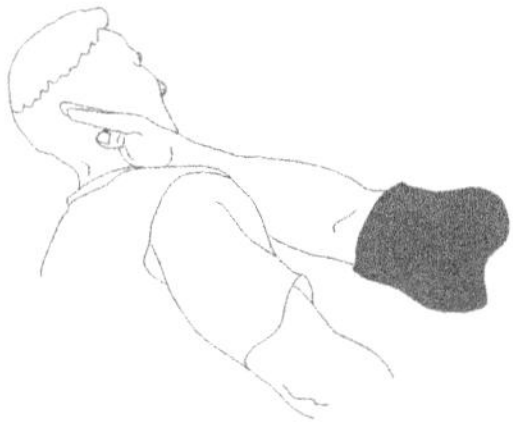

Vertical Knife Hand. When thrown vertically, the knife hand strike comes straight down in a straight line.

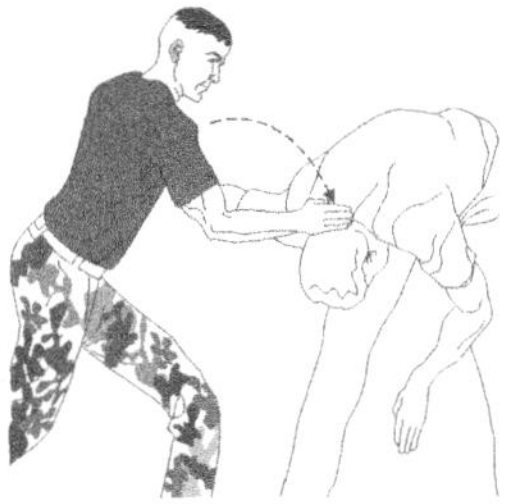

Eye Gouge

The eye gouge is used to attack an opponent's eyes, blinding him so follow-up strikes can be executed. The striking surface is the tips of the fingers or thumb. To execute the eye gouge, Marines—

Extend the right hand with the fingers slightly spread apart to allow entry into the eye sockets.

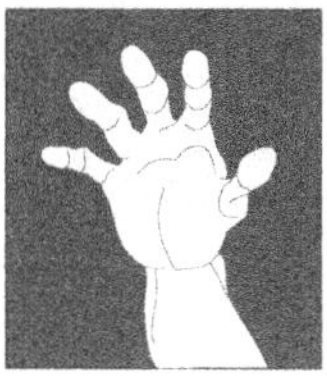

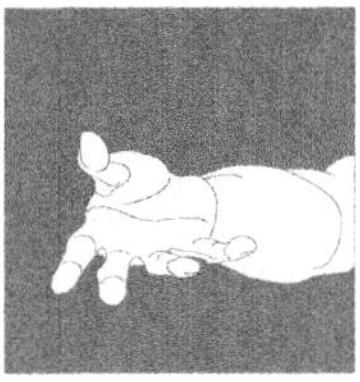

Place the palm of the hand either toward the ground or toward the sky and thrust the right hand forward into the opponent's eyes.

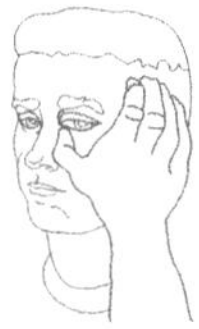

Thrust the hand forward at the opponent's nose level so the fingers or thumb slide naturally into the grooves of the opponent's eye sockets.

Elbow Strikes

The elbow is a powerful weapon that can be used in several different ways to attack virtually any part of an opponent's body. Elbow strikes can be performed either vertically (upward or downward) or horizontally (forward or reverse). The striking surface is 2 inches above or below the point of the elbow, depending upon the angle of attack, the opponent's attack angle, and the position of the opponent.

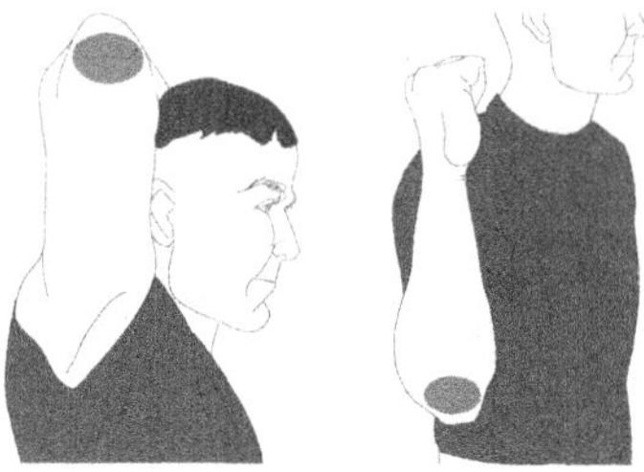

Above Elbow **Below Elbow**

Vertical Elbow Strike (Up). To execute an upward vertical elbow strike, Marines—

Bend the right elbow, keeping the fist close to the body. The fist is at shoulder level and the elbow is next to the torso.

Thrust the elbow vertically upward toward the opponent while rotating the right shoulder and hip forward to generate additional power.

Contact the opponent with the right forearm 2 inches above the point of the elbow.

Vertical Elbow Strike (Down). To execute a downward vertical elbow strike, Marines—

Bend the right elbow, keeping the fist close to the body. The fist is on the shoulder and the elbow is raised well above the shoulder.

Thrust the elbow vertically downward toward the opponent while dropping body weight into the attack to generate additional power.

Contact the opponent with the right triceps 2 inches above the point of the elbow.

Horizontal Elbow Strike (Forward). To execute the forward horizontal elbow strike, Marines—

Tuck the right fist near the chest with the palm heel facing the ground.

Thrust the right elbow horizontally forward toward the opponent. The forearm is parallel to the ground.

Rotate the right shoulder and hip forward to generate additional power.

Contact the opponent with the right forearm 2 inches below the point of the elbow.

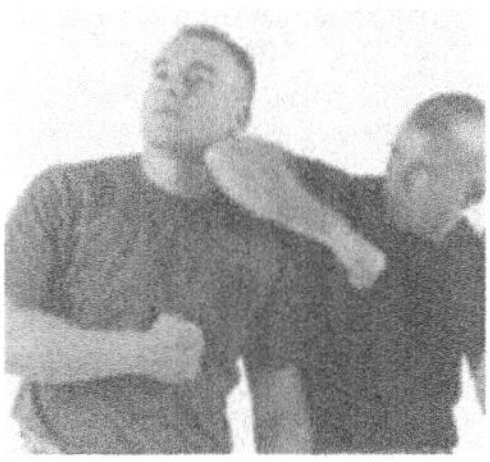

Horizontal Elbow Strike (Rear). To execute the rear horizontal elbow strike, Marines—

Tuck the right fist near the left shoulder with the palm heel facing the ground. At the same time, rotate the right shoulder forward and the left hip forward.

Thrust the right elbow horizontally rearward toward the opponent. The forearm is parallel to the ground and the hand moves toward the direction of the attack.

Rotate the right hip back and the right shoulder backward to generate additional power.

Contact the opponent with the right triceps 2 inches above the point of the elbow.

4. Strikes with the Lower Body

The legs are the body's most powerful weapons because they use the largest muscle groups to generate a strike. Legs are also less prone to injury. The feet are the preferred choice for striking because they are protected by boots. Marines use their feet, heels, and knees to execute kicks, knee strikes, and stomps.

Kicks

The purpose of a kick is to stop an opponent's attack or to create an opening in his defense in order to launch an attack. Kicks can be performed with the left (lead) leg or the right (rear) leg. Kicks with the rear leg have greater power because the hips are rotated into the attack. However, the rear leg is further away from the opponent so a strike with the rear leg will not contact the opponent as quickly as a strike with the lead leg.

Front Kick. The front kick is executed when the opponent is in front of the Marine. The front kick, delivered with either the rear or lead leg, is effective for striking below the waist. Attempting to kick higher results in diminished balance and provides the enemy with a greater opportunity to grab the leg or foot. The striking surfaces are the toe of the boot or the bootlaces. To execute the front kick, Marines—

Raise the left knee waist high, pivot the hips into the attack, and thrust the left foot forward toward the opponent.

Contact the opponent with the toe of the left boot or bootlaces.

Return to the basic warrior stance.

Side Kick. The side kick, delivered with the lead leg, is effective for striking the knees. The side kick is executed when the opponent is to the side of the Marine. The striking surface is the outside cutting edge of the boot near the heel. To execute the side kick, Marines—

Raise the right knee waist high and rotate the right hip forward.

Thrust the right foot to the right side toward the opponent, turning the foot at a 90-degree angle to maximize the striking surface on the opponent.

Contact the opponent with the cutting edge of the right boot.

Return to the basic warrior stance.

Knee Strikes

Knee strikes are excellent weapons during the close range of close combat fighting. The knee strike is generally delivered in close.

Vertical Knee Strike. The striking surface is the thigh, slightly above the knee. To execute the vertical knee strike, Marines—

Raise the right knee and drive it up forcefully into the opponent. Power is generated by thrusting the leg upward.

Contact the opponent 2 inches above the right knee.

Horizontal Knee Strike. The horizontal knee strike is executed with the leg generally parallel with the ground while rotating the hips to generate power. It is often delivered to the peroneal nerve. The striking surface is the front of the leg, slightly above or below the knee. To execute the horizontal knee strike, Marines—

Raise the right knee, rotate the right hip forward while pivoting on the left foot, and drive the knee horizontally into the opponent.

Contact the opponent 2 inches above the right knee.

Stomps

Stomps are delivered with the feet, usually when the opponent is down. Remember, when the opponent is down, Marines take whatever target is available.

Vertical Stomp. The vertical stomp allows Marines to remain upright and balanced, to rapidly deliver multiple blows with either foot, and to quickly and accurately attack the target. It is the preferred stomp. The striking surface is the flat bottom of the boot or the cutting edge of the heel. To execute the vertical stomp, Marines—

Raise the right knee above the waist with the right leg bent at approximately a 90-degree angle.

Drive the flat bottom of the right boot or the cutting edge of the right heel down onto the opponent forcefully. At the same time, bend the left knee slightly to drop the body weight into the strike.

Ax Stomp. The striking surface of the ax stomp is the cutting edge of the heel. To execute the ax stomp, Marines—

Raise the right heel above the waist, keeping the right leg straight.

Drive the cutting edge of the right heel down onto the opponent forcefully. At the same time, bend the left knee slightly to drop the body weight into the strike.

5. Counters to Strikes

In a close combat situation, an opponent will attempt to strike Marines with punches and kicks. When an opponent uses a strike, Marines must first avoid the strike. This is accomplished by moving quickly and blocking. Next, Marines must get into an offensive position. This allows Marines to use offensive strikes to attack the opponent. Regardless of the strike, the counter to a strike requires Marines to move, block, and strike.

Move

The first step in countering a strike is to move out of the way of the strike's impact. Movement removes Marines from the opponent's intended strike point and positions Marines to attack. Movement is executed at approximately a 45-degree angle to the front or rear. Movement is always initiated from the basic warrior stance. Return to the basic warrior stance with the toe of the lead foot pointing toward the opponent once the movement is complete.

Block

Different blocks are executed based on the strike. These will be covered by individual counters.

Strike

Any of the upper body or lower body strikes can be executed as a follow-on attack or part of the counter to an opponent's strike. The follow-on strike is determined by the angle to the opponent, the position of the opponent, and the opponent's available vulnerable target areas.

Counters to Punches

Counter to a Lead Hand Punch. This counter is used when the opponent throws a lead hand punch. To execute the counter to the lead hand punch, Marines—

Step forward and to the left at approximately a 45-degree angle, moving in to the outside of the opponent's attacking arm.

Raise the left arm and block or deflect the opponent's lead hand with the palm of the hand or the meaty portion of the forearm.

Hit and stick by leaving the left arm against the opponent's right arm while stepping forward and to the left at approximately a 45-degree angle to close with the opponent.

Execute a finishing technique, such as a strike or a kick, to the opponent's exposed target areas.

Counter to a Rear Hand Punch. This counter is used when the opponent throws a rear hand punch. To execute the counter to the rear hand punch, Marines—

Step forward and to the left at approximately a 45-degree angle, moving in to the outside of the opponent's attacking arm.

Raise the left arm and block or deflect the opponent's rear hand with the palm of the hand or the meaty portion of the forearm.

Hit and stick by leaving the left arm against the opponent's right arm while stepping forward and to the right at approximately a 45-degree angle to close with the opponent.

Execute a finishing technique, such as a strike or a kick, to the opponent's exposed target areas.

Counters to Kicks

Counter to a Front Kick (Left or Lead Leg). This counter is used when the opponent executes a front kick with his left leg. To execute the counter to a front kick, Marines—

Step forward and to the right at approximately a 45-degree angle, moving in to the outside of the opponent's striking leg.

Lower the left arm and block or deflect the opponent's leg with the palm of the hand or the meaty portion of the forearm.

Hit and stick by leaving the left arm against the opponent's leg while stepping forward and to the left at approximately a 45-degree angle to close with the opponent.

Execute a finishing technique, such as a strike or a kick, to the opponent's exposed target areas.

Counter to a Front Kick (Right or Rear Leg). This counter is used when the opponent executes a front kick with his right leg. To execute the counter to a front kick, Marines—

Step forward and to the left at approximately a 45-degree angle, moving in to the outside of the opponent's striking leg.

Lower the left arm and block or deflect the opponent's leg with the palm of the hand or the meaty portion of the forearm.

Hit and stick by leaving the left arm against the opponent's leg while stepping forward and to the right at approximately a 45-degree angle to close with the opponent.

CHAPTER 5

THROWS

This chapter describes all techniques for a right-handed person. However, all techniques can be executed from either side.

The Marine is depicted in camouflage utilities. The opponent is depicted without camouflage.

Marines use throwing techniques to maintain the tactical advantage and to throw the opponent to the ground during close combat. Throws apply the principles of balance, leverage, timing, and body position to upset an opponent's balance and to gain control by forcing the opponent to the ground. Throwing techniques are effective because they are size- and gender-neutral, and they rely on the momentum and power generated by the opponent rather than the strength or size of the Marine. Marines also execute a throw as a devastating attack against an opponent, possibly causing unconsciousness or broken limbs. When Marines execute throws, they must maintain balance and, simultaneously, prevent the opponent from countering a throw or escaping after being forced to the ground.

1. Turning Throw

Marines use a turning throw to take the opponent to the ground while they remain standing. A turning throw can also be executed from a stationary position. It is particularly effective if Marines and the opponent are wearing gear. To execute the turning throw, Marines—

Grasp the opponent's right wrist with the left hand.

Step forward with the right foot, place it against the outside of the opponent's right foot, and pivot so the back of the heel is next to the middle of the opponent's foot.

Hook the opponent's right arm with the right arm and pinch his arm between the biceps and forearm, touching the opponent with the body.

Pull the opponent's wrist downward, keeping it close to the body.

Pivot to the left on the ball of the foot and continue pulling downward on the opponent's wrist while rotating the wrist outward to off-balance him.

2. Hip Throw

Marines use a hip throw to take the opponent to the ground while they remain standing. A hip throw is particularly effective if the opponent is moving forward or pushing. Marines use the opponent's forward momentum to execute the hip throw. To execute the hip throw, Marines—

Grasp the opponent's right wrist with the left hand.

Step forward with the right foot and place it against the outside of the opponent's right foot.

The back of the heel should be next to the opponent's foot.

Step back with the left foot and rotate on the ball of the foot. The back of the heel is next to the opponent's toe. The knees are bent.

Rotate at the waist and hook the right hand around the back of the opponent's body (anywhere from his waist to his head). The side and hip should be against the opponent.

Rotate the hip against the opponent. The hips must be lower than the opponent's.

Use the right hand to pull the opponent up on the hip to maximize contact.

Pull the opponent's arm across the body and, at the same time, lift the opponent off the ground slightly by bending at the waist, straightening the legs, and rotating the body to the left.

3. Leg Sweep

Marines use a leg sweep to take the opponent to the ground while they remain standing. A leg sweep is particularly effective if the opponent is already off-balanced and moving backward or pulling on the Marine. To execute the leg sweep, Marines—

Grab the opponent's right wrist with the left hand and grab the opponent's left shoulder with the right hand.

Note: Grab the opponent's clothing or gear if his wrist and shoulder cannot be grabbed.

Step forward with the left foot and place it on the outside of the opponent's right foot.

Pull the opponent's wrist downward, close to the body, and push his shoulder backward to off-balance him.

Raise the right knee no higher than the waist.

Kick the foot past the opponent's right leg.

Use the heel of the boot to make contact with the opponent's calf (anywhere from the top of the calf down to the Achilles tendon or on the inside of the calf).

Sweep the opponent to the ground.

CHAPTER 6

CHOKES AND HOLDS

This chapter describes all techniques for a right-handed person. However, all techniques can be executed from either side.

The Marine is depicted in camouflage utilities. The opponent is depicted without camouflage.

WARNING

During training, never execute a choke at full force or full speed and never hold a choke for more than 5 seconds.

When Marines correctly perform a choke, they render an opponent unconscious in as little as 8 to 13 seconds. Chokes are easily performed regardless of size or gender. Marines must know how to apply chokes and how to counter a choke or a hold executed by an opponent.

1. Types of Chokes

There are two types of chokes: an air choke and a blood choke. An air choke closes off the airway to the lungs, thereby preventing oxygen from reaching the heart. A blood choke cuts off the blood flow to the brain. Both types can result in unconsciousness and eventual death for an opponent.

Air Choke

An air choke is performed on the opponent's windpipe (or trachea), cutting off the air to the lungs and heart. If Marines execute the air choke properly, the opponent loses consciousness within 2 to 3 minutes. Due to the length of time it takes to immobilize the opponent, air chokes are not recommended.

Blood Choke

A blood choke is performed on the opponent's carotid artery, which carries oxygen-enriched blood from the heart to the brain. The carotid artery is located on both sides of the neck. If Marines execute a blood choke properly, the opponent will lose consciousness within 8 to 13 seconds. The blood choke is the preferred choke because its intended effect (i.e., the opponent losing consciousness) can be executed quickly, ending the fight.

2. Chokes

Front Choke

Marines execute a front choke when they are facing the opponent. Marines use the opponent's lapels or collar to execute a front choke. To execute the front choke, Marines—

Grab the opponent's right lapel with the right hand, making certain that the knuckles or the back of the hand are against the opponent's carotid artery on the right side of his neck.

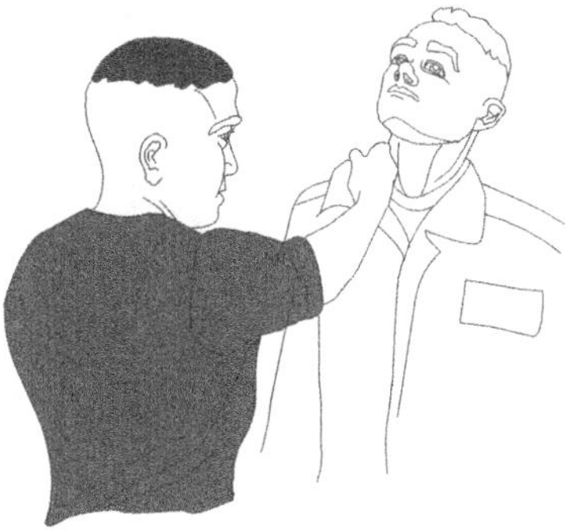

Keep the right hand against the opponent's neck, reach under the opponent's right arm with the left hand, grab the opponent's left lapel, and form an X with the wrists.

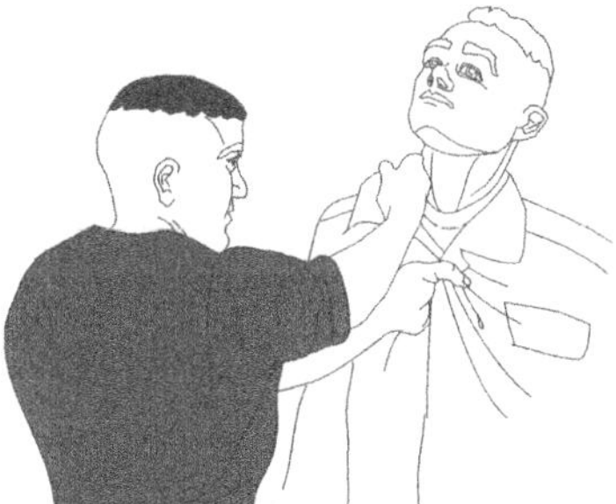

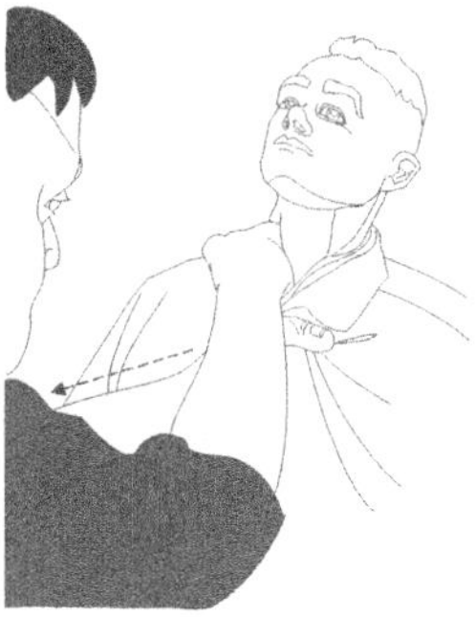

Press the right hand against the opponent's carotid artery.

Pull his lapel to the left with the left hand.

Side Choke

Marines execute a side choke when they are facing the opponent. The side choke is particularly effective when deflecting a punch thrown by an opponent. To execute the side choke, Marines—

Use the left hand to parry the opponent's arm inboard (to the inside of the opponent's reach).

Bring the right arm underneath the opponent's arm and up around the front of his neck.

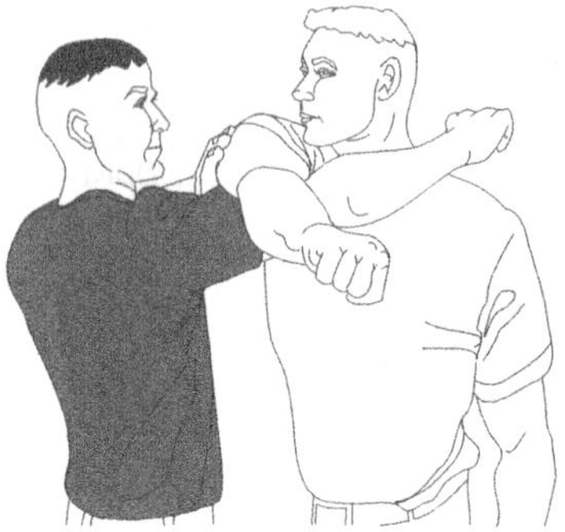

Extend the fingers, place the back of the forearm against the opponent's neck just below his ear, and press the carotid artery on his neck.

Note: If Marines are unable to place their fingers against the opponent's neck, they may use the back of the thumb or wrist.

Reach, with the left hand, around the back of the opponent's neck and clasp the hands together.

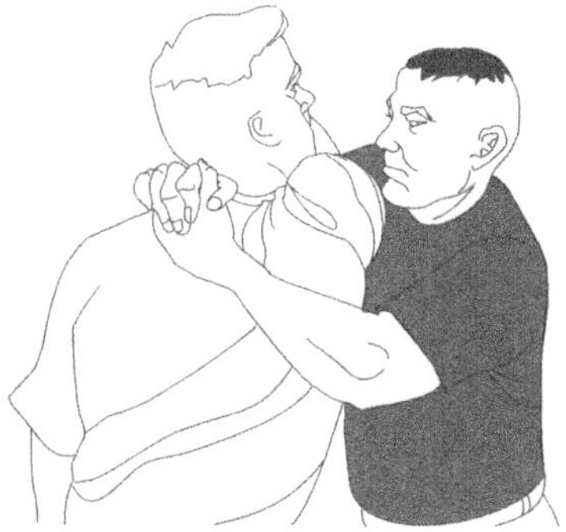

Pull the opponent toward the chest by pulling the clasped hands toward the chest.

Exert pressure on the side of the opponent's neck with the forearm.

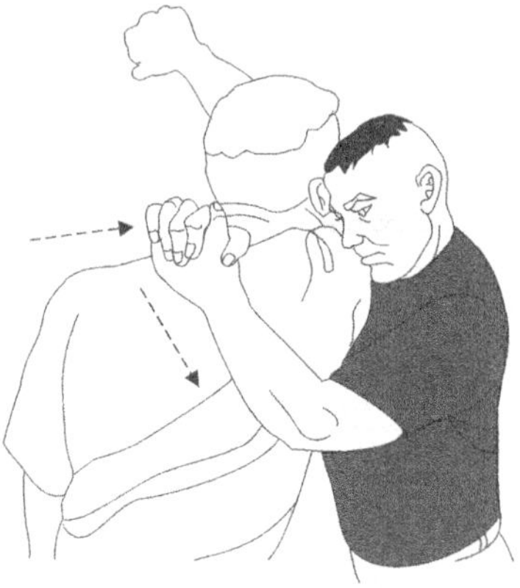

Rear Choke

Marines execute a rear choke when they are behind the opponent, the opponent is on the ground, or when they are taking the opponent to the ground. To execute the rear choke, Marines—

Reach, with the right arm, over the opponent's right shoulder and hook the bend of the arm around his neck.

Clasp both hands together.

Exert pressure with the biceps and forearms on both sides of the opponent's neck on his carotid arteries.

Maintain pressure with the biceps and forearms on both sides of the neck and draw the opponent closer by drawing the right arm in.

Figure-Four Choke

A variation of the rear choke is the figure-four choke. The figure-four choke allows Marines to gain more leverage than the rear choke. If the rear choke cannot be secured, the figure-four variation is employed to increase the pressure of the choke on the opponent. To execute the figure-four variation of the rear choke, Marines—

Apply a rear choke. The Marine's body should be against the opponent's body.

Grasp the left biceps with the right hand and place the left hand against the back of the opponent's head.

Push the opponent's head forward and down with the left hand.

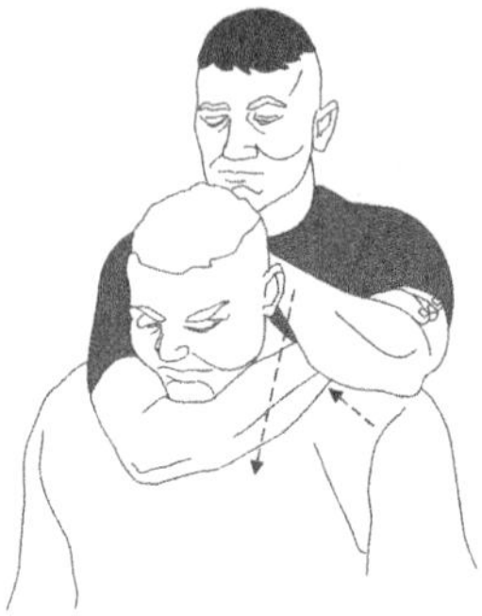

Draw the right arm in, maintaining pressure with the bicep and forearm on both sides of the opponent's neck.

3. Counters to Chokes and Holds

During a close combat situation, an opponent may apply a choke or hold on a Marine. If the opponent correctly applies a choke, a Marine quickly loses consciousness. If a choke is not executed properly, it often results in a hold, typically a bear hug or a headlock. A hold allows the opponent to control a Marine and removes the Marine's ability to attack. It is important for Marines to extract themselves from chokes and holds, regain the tactical advantage, and counter with strikes.

Although a choke causes unconsciousness in 8 to 13 seconds for a blood choke and 2 to 3 minutes for an air choke, the first movement in any counter to a choke is to clear the airway. Marines use softening techniques to loosen an opponent's grip and to clear their airway. Softening techniques are particularly effective if Marines lack the physical strength of their opponent. These techniques include groin strikes, eye gouges, foot stomps, etc. Softening techniques are not offensive; rather, they are used to loosen an opponent's hold.

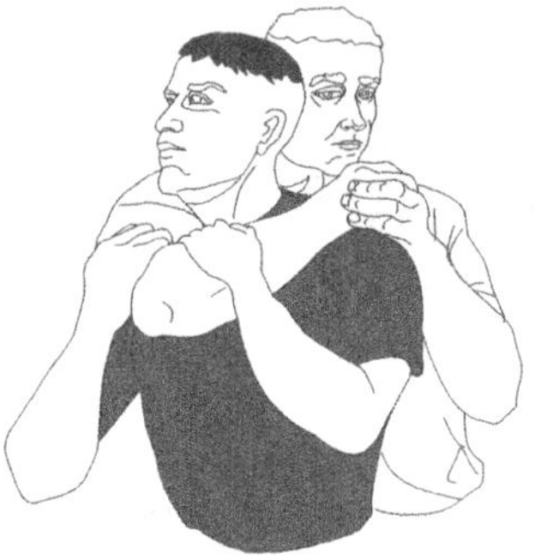

The second movement is to tuck the chin. Once the airway is clear, Marines tuck their chins to prevent the opponent from reapplying the choke.

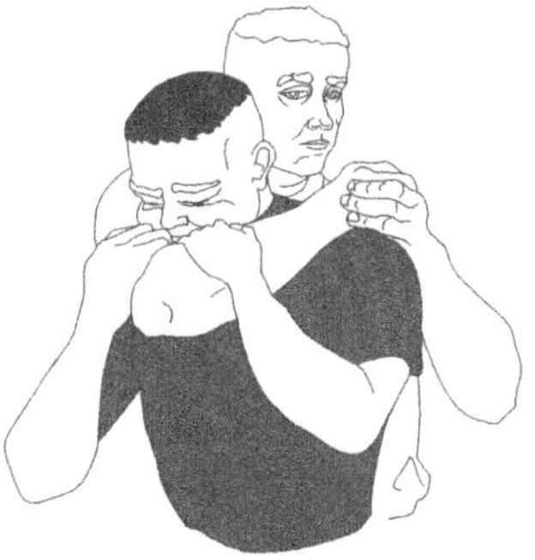

Counter to a Front Choke

Marines use a counter to a front choke when the opponent approaches from the front and uses both hands to choke a Marine around the throat. To execute the counter to the front choke, Marines—

Grasp the opponent's right forearm (where the elbow bends) with the left hand and apply downward pressure on the opponent's radial nerve with the fingers.

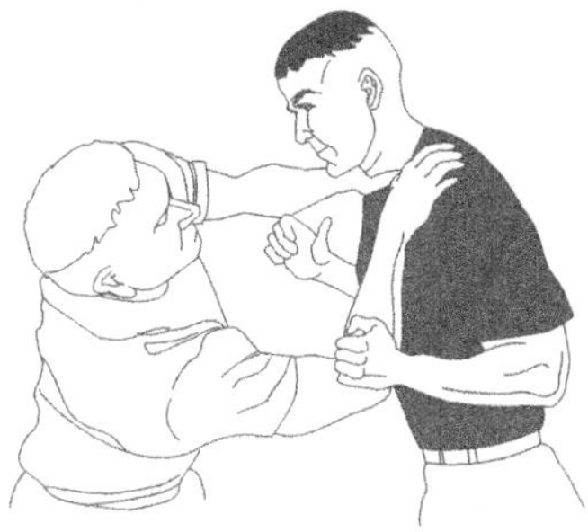

Execute a chin jab to the opponent's chin with the right hand. To generate power into the strike, bring the left foot to the outside of the opponent's right foot and rotate the hips into the strike.

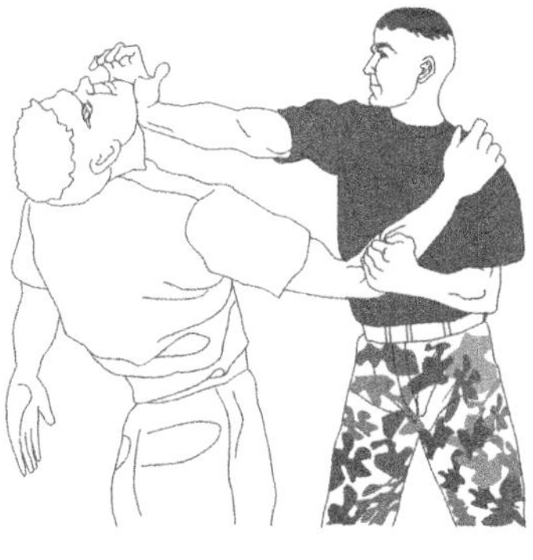

Counter to a Rear Choke

Marines execute a counter to a rear choke when the opponent approaches from the rear and puts his right arm around a Marine's throat. To execute the counter to the rear choke, Marines—

Grasp the opponent's forearm (at the radial nerve) and bicep with both hands and pull down just enough to clear the airway. Once the airway is clear, tuck the chin to protect the airway and to prevent the opponent from reapplying the choke.

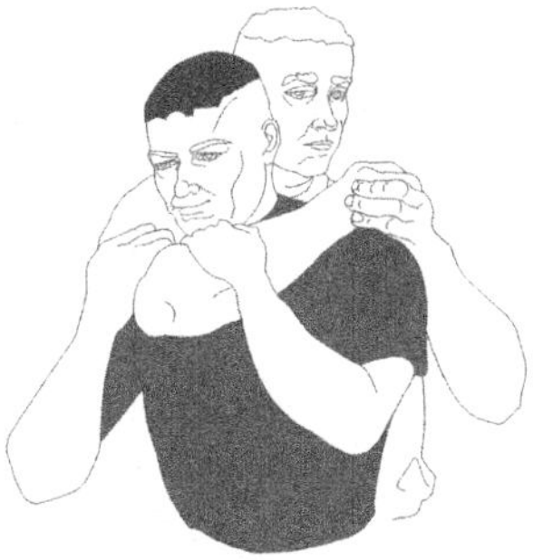

Step behind the opponent's right leg with the left foot, keeping both legs bent (almost in a squatting position).

Strike and drive the left elbow into the opponent's torso while rotating the hips and pivoting to the left, throwing the opponent back and to the ground.

Counter to a Front Headlock

Marines use a counter to a front headlock when the opponent approaches from the front and puts his right arm around the Marine's neck, bends the Marine forward, and locks the Marine's head against his hip. To execute the counter to a front headlock, Marines—

Grasp the opponent's wrist and forearm with both hands and pull down to clear the airway. Maintain control of the opponent's wrist throughout the move. Once the airway is clear, tuck the chin to protect the airway and to prevent the opponent from reapplying the choke.

Move the right hand and arm across the opponent's torso.

Step forward and to the left with the left foot at a 45-degree angle.

Execute a sweep with the right foot against the opponent's right leg. At the same time, push against the opponent's chest with the right arm and shoulder to generate power in the sweep.

Counter to a Rear Headlock

Marines use a counter to a rear headlock when the opponent approaches from the rear and puts his right arm around the Marine's neck, bends the Marine forward, and locks the Marine's head against his hip. To execute the counter to the rear headlock, Marines—

Grasp the opponent's wrist and forearm with the right hand and pull down to clear the airway. Once the airway is clear, tuck the chin to protect the airway and to prevent the opponent from re-applying the choke.

Reach over the opponent's right shoulder with the left arm.

Grab any part of the opponent's face (chin, nose, eyes) and pull back while rising to a standing position.

Execute, with the right hand, a hammer fist strike to the opponent's exposed throat.

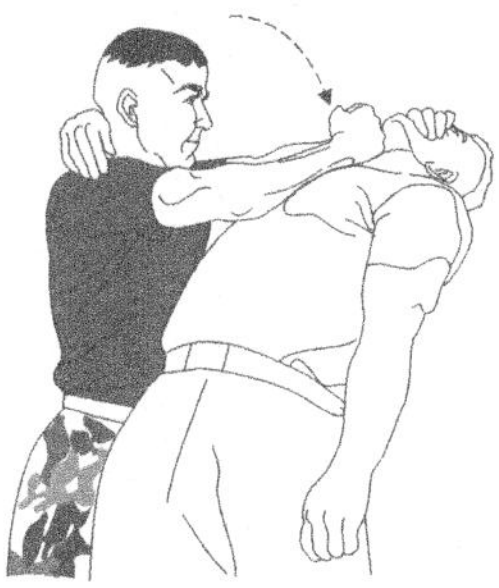

Counter to a Front Bear Hug

Marines execute a counter to a front bear hug when the opponent approaches from the front and puts both of his arms around the Marine's body, trapping the Marine's arms to the sides. To execute the counter to a front bear hug, Marines—

Step forward and to the left with the left foot at a 45-degree angle to the outside of the opponent's right leg, keeping the left leg bent.

Grasp the opponent's torso or arms to gain balance and to assist in throwing the opponent. It may be helpful to hook the opponent's right arm with the left arm.

Drive the right arm and shoulder forward and, at the same time, bring the right leg forward and sweep the opponent's right leg, bringing him to the ground.

Counter to a Rear Bear Hug

Marines use a counter to a rear bear hug when the opponent approaches from the rear and puts both of his arms around the Marine's body, trapping the Marine's arms to the sides. To execute the counter to a rear bear hug, Marines—

Step behind the opponent's right leg with the left foot, keeping both legs bent (almost in a squatting position). The left side of the body should be against the opponent's.

Pivot the hip, turning the body to the left and throwing the opponent back over the bent leg.

CHAPTER 7

GROUND FIGHTING

This chapter describes all techniques for a right-handed person. However, all techniques can be executed from either side.

In drawings, the Marine is depicted in woodland camouflage utilities; the opponent is depicted without camouflage. In photographs, the Marine is depicted in woodland camouflage utilities; the opponent is depicted in desert camouflage utilities.

Marines should avoid being on the ground during a close combat situation because the battlefield may be covered with debris and there is an increased risk of injury. However, many close combat situations involve fighting on the ground. The priority in a ground fight is for Marines to get back on their feet as quickly as possible. In any ground fighting scenario, Marines will usually end up in one of four positions with the opponent. The offensive positions, in which Marines have a tactical advantage, are the guard and mount. The defensive positions, which are used as counters when the opponent has the tactical advantage, are the counter to the guard and the counter to the mount. Marines can also employ chokes during ground fighting to quickly end a fight.

1. Offensive Ground Fighting

Guard Position

Marines execute a guard position when the opponent is on top and the legs are wrapped around the opponent's legs. If the opponent is on top, he may try to choke the Marine, but the Marine still has the tactical advantage because the Marine is in a position to control the situation. To execute the guard position technique, Marines—

Trap the opponent's hands on the chest by crossing the hands on the chest so the forearms are resting on the opponent's forearms. Apply pressure with the elbows or forearms.

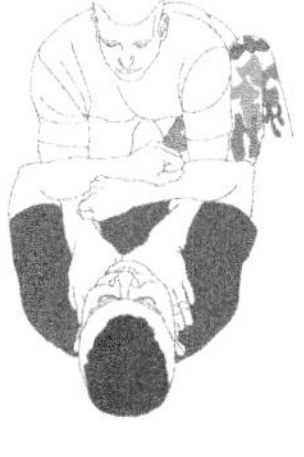

Strike the outside of the opponent's thigh with the cutting edge of the right heel. This causes the opponent to jerk to that side.

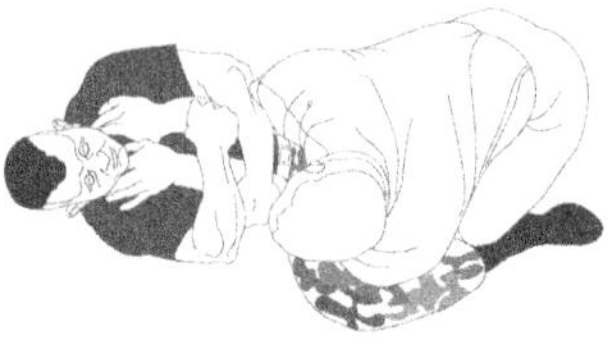

Move the head quickly to the left and swivel the hips to the right. At the same time, bring up both of the legs. Both legs are on the right side of the opponent's body.

Bring the right leg down, hooking the opponent's neck and head. Exert downward pressure to roll him over on his back. Grasp and maintain control of the opponent's left arm. Upon completion of the move, the Marine is sitting up with legs bent over the opponent while maintaining control of his arm.

Keep legs and knees bent. Maintain pressure against the opponent's neck with the back of the right foot and against his side with the left foot underneath his armpit. Squeeze the knees together, locking the opponent's arm.

Pull the opponent's arm straight up and fall back sharply, pulling his arm to the side in the direction of his little finger.

Return to standing.

Mount Position

Marines execute the mount if the opponent is lying on his back on the ground and the Marine is on top with legs wrapped around the opponent's body. This position is an offensive position because Marines are in a better position to control the opponent and to execute ground fighting techniques. To execute the mount position technique, Marines—

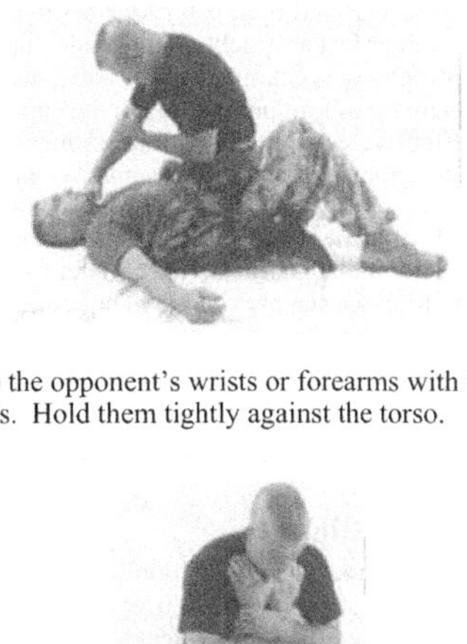

Grab the opponent's wrists or forearms with the hands. Hold them tightly against the torso.

Maintain control of the opponent's arms with the left arm and apply pressure to the opponent's brachial plexus (tie in) with the right hand.

Bring the left foot close to the opponent's right armpit and swing the right leg across the opponent's head.

Keep the legs and knees bent. Maintain pressure against the opponent's neck with the back of the right foot and against his side with the left foot which is underneath his armpit. Squeeze the knees together, locking the opponent's arm.

Pull the opponent's arm straight up and fall back sharply, pulling his arm to the side in the direction of his little finger.

Return to standing.

2. Defensive Ground Fighting

Counter to the Guard

Marines use this technique if the opponent is lying on his back on the ground and the Marine is kneeling on the ground between the opponent's legs. To execute the counter to the guard position, Marines—

Strike the opponent's femoral nerve, located on the inside of the thigh, with the elbows. This forces the opponent to separate his legs.

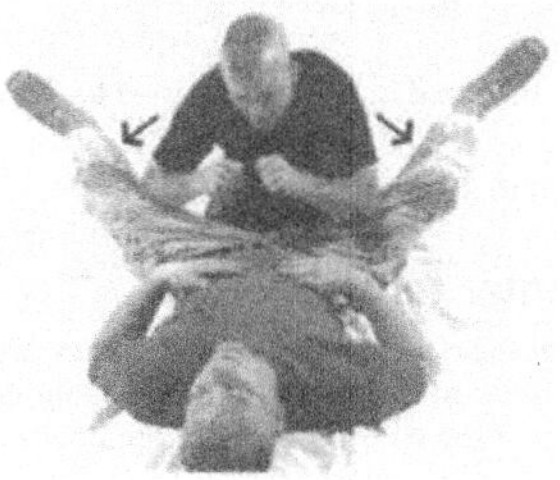

Strike the opponent's groin with the right fist.

Hook the left arm underneath the opponent's right knee from the inside. Throw the opponent's leg over the head.

Duck the head quickly and move to the left.

Return to standing.

Counter to the Mount

Marines use this technique when they are lying on their back on the ground, the opponent is mounted on top of the Marine, and the opponent's legs are wrapped around the Marine. The opponent has the tactical advantage. To execute the counter to the mount position, Marines—

Grab the opponent's gear or clothing on his upper torso and pull him down close.

Use the right arm to hook the opponent's left arm, from the inside around the outside, above his elbow. With the right foot, hook the opponent's left leg or ankle.

Draw the elbow in to bend the opponent's elbow, bringing him down close. The arm must be hooked above the opponent's elbow in order to bend it.

Strike the opponent's side with the left hand.

Push the opponent over and roll him off to the right side.

Maintain control of the opponent's hooked arm and move to a kneeling position.

Move to a standing position while maintaining pressure on the opponent's arm and using the knee to apply pressure against the opponent's elbow.

3. Ground Fighting Chokes

WARNING

During training, never execute a choke at full force or full speed. Never hold a choke for more than 5 seconds.

The priority in ground fighting is for Marines to get back on their feet as quickly as possible. Sometimes, Marines can quickly end a ground fight by executing a choke on the opponent. When performed correctly, a choke can render an opponent unconscious in as little as 8 to 13 seconds. Chokes are easily performed regardless of size or gender. The chokes performed during ground fighting are the same as those performed while standing (see chap. 6).

Ground Fighting Front Choke

The ground fighting front choke is a blood choke performed most effectively from the mount position. The front choke employs the opponent's lapels or collar to execute the choke. To execute the ground fighting front choke, Marines—

Grab the opponent's right lapel with the right hand, making certain that the knuckles or the back of the hand are against the carotid artery on the right side of the opponent's neck.

Keep the right hand pressed against the opponent's carotid artery, reach under the right arm with the left hand, and grab the opponent's left lapel, forming an X with the wrists.

Keep the right hand pressed against the opponent's carotid artery and pull the opponent's left lapel to the left with the left hand.

Ground Fighting Side Choke

The ground fighting side choke is a blood choke performed from the mount position. The ground fighting side choke is particularly effective when the opponent raises his arms and places them on the Marine's chest or throat. To execute the ground fighting side choke, Marines—

Use the left hand to parry the opponent's right arm inboard (to the inside of the opponent's reach). Bring the right arm underneath the opponent's arm and up around the front of his neck.

Keep the fingers extended, place the back of the forearm against the opponent's neck just below his ear and press the carotid artery on his neck. Depending on the Marine's or the opponent's body configuration, the back of the thumb or wrist may press against the opponent's neck.

Use the left hand to reach around the back of the opponent's neck and clasp the hands together.

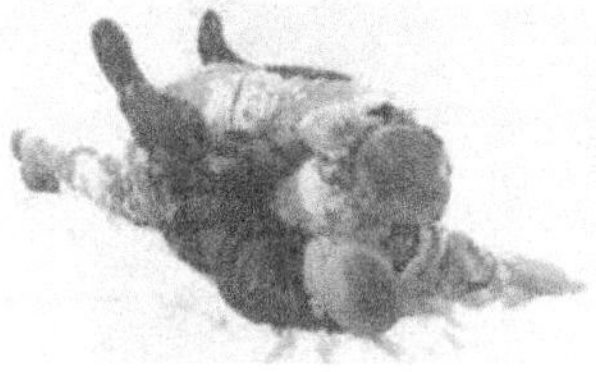

Pull the opponent toward the chest. Use the fore-arm to exert pressure on the side of his neck. This is done by pulling the clasped hands toward the chest.

Ground Fighting Rear Choke

The ground fighting rear choke is a blood choke performed when Marines are behind the opponent. To execute the ground fighting rear choke, Marines—

Wrap the legs around the opponent with the left foot against the inside of his left thigh and the right foot against the inside of his right thigh.

Use the right arm to reach over the opponent's right shoulder and hook the bend of the arm around his neck.

Use the left hand to clasp both hands together.

Exert pressure with the biceps and forearms on both sides of the opponent's neck. This applies pressure to the opponent's carotid arteries. While maintaining pressure with the biceps and fore-arms, draw the opponent closer by drawing the right arm in. To increase the effectiveness of the choke, lean backward by arching the back and pulling the opponent back. At the same time, push the feet against the opponent's thighs.

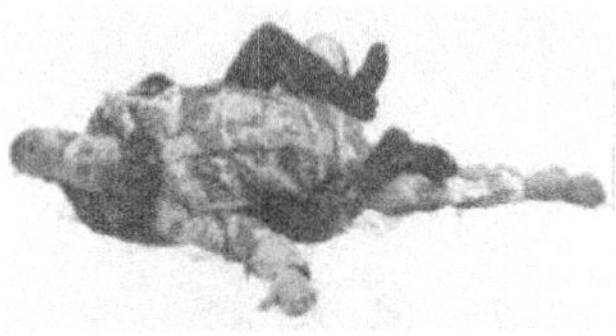

Figure-Four Choke

A variation of the ground fighting rear choke is the figure-four choke. The figure-four choke allows Marines to gain more leverage on the rear choke. If the rear choke cannot be secured, the figure-four variation may be applied to increase the pressure of the choke on the opponent. To execute the figure-four variation of the rear choke, Marines—

Apply the ground fighting rear choke. The chest should be against the opponent's back.

Grasp the left biceps with the right hand and place the left hand against the back of the opponent's head.

Use the left hand to push the opponent's head forward and down.

Draw the right arm in while maintaining pressure with the biceps and forearm on both sides of the opponent's neck.

Increase the effectiveness of the choke by arching backward and pushing the feet against the opponent's thighs. Continue to pull the opponent back with the right arm while exerting pressure forward with the left hand against the opponent's head.

CHAPTER 8

NONLETHAL TECHNIQUES

This chapter describes all techniques for a right-handed person. However, all techniques can be executed from either side.

In drawings, the Marine is depicted in woodland camouflage utilities; the opponent is depicted without camouflage. In photographs, the Marine is depicted in woodland camouflage utilities; the opponent is depicted in desert camouflage utilities.

The Marine Corps' involvement in military operations other than war—e.g., humanitarian, peacekeeping, or evacuation missions—has greatly increased. These missions require skills that span the spectrum of conflict and support operations within a continuum of force. But the Corps' day-to-day existence also demands a responsible use of force. Nonlethal techniques are among the skills Marines use to apply a responsible use of force.

1. Unarmed Restraints and Manipulation

Marines operate within a continuum of force, particularly in support of peacekeeping or humanitarian types of missions. In these situations, Marines must act responsibly to handle a situation without resorting to deadly force. Unarmed restraints and manipulation techniques including joint manipulation, come-alongs, and takedowns can be used to control a subject without resorting to deadly force. Marines must train to become proficient in nonlethal techniques and to respond in a responsible manner. These techniques are referred to as compliance techniques, and they are applied in the third level in the continuum of force.

WARNING

During training, never apply the techniques for unarmed restraints and manipulation at full force or full speed. Use a slow and steady pressure to avoid injury.

Compliance Techniques

Compliance techniques are unarmed restraint and manipulation techniques used to physically force a subject or opponent to comply. Compliance can be achieved through the close combat techniques of—

- Pain compliance using joint manipulation and pressure points. (Pain compliance is the initiation of pain to get compliance on the part of the subject.)
- Come-along holds.

Principles of Joint Manipulation

Joint manipulation is used to initiate pain compliance and gain control of a subject. It involves the application of pressure on the joints (elbow, wrist, shoulder, knee, ankle, and fingers). Pressure is applied in two ways:

- In the direction in which the joint will not bend. For example, joints such as the knees and elbows only bend in one direction and when pressure is applied in the opposite direction, pain compliance can be achieved.

- Beyond the point where the joint stops naturally in its range of movement (i.e., it no longer bends).

Since each joint has a breaking point, Marines should apply slow steady pressure only until pain compliance is reached. Continued pressure will break the joint and may escalate the violence of the situation.

Joint manipulation also uses the principle of off-balancing. A subject can be better controlled when he is knocked off balance.

Wristlocks

A wristlock is a joint manipulation that can be applied in a number of ways to achieve pain compliance. The wrist rotates in a number of directions and will bend in a single direction until its movement stops naturally. In a wristlock, pressure is exerted beyond that point by bending or twisting the joint. A wristlock is executed when an opponent tries to grab Marines or is successful in grabbing Marines or their equipment. A wristlock can also be performed by Marines if they wish to initiate control of an opponent.

Basic Wristlock. A basic wristlock is executed when Marines grab the opponent's left hand with the right hand. To execute the basic wristlock, Marines—

Use the right hand to grab the opponent's left hand by placing the thumb on the back of the opponent's hand so that the Marine's knuckles are facing to the left.

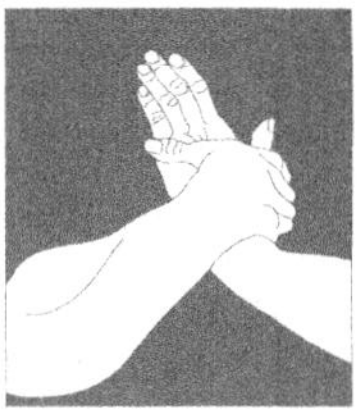

Hook the fingers across the fleshy part of the opponent's palm below the thumb. The fingers are used to anchor the hand so leverage can be applied to twist and bend the joint.

Exert downward pressure with the thumb to bend the opponent's joint. Rotate the opponent's hand to the right to twist the joint.

Step in to the opponent to keep the opponent's hand in close to the body to control him and provide more leverage on the wristlock.

Note: When executing the basic wristlock with the left hand, the Marine grabs the opponent's hand so that the Marine's knuckles are facing to the right, and then rotates and twists the opponent's hand to the left.

Use the left hand to further control the opponent.

Reverse Wristlock. A reverse wristlock is executed when Marines grab the opponent's right hand with the right hand. To execute the reverse wristlock, Marines—

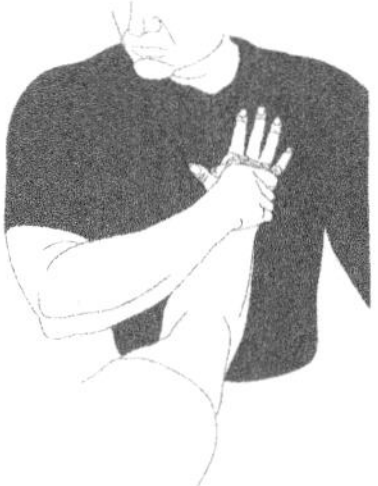

Place the right palm on the back of the opponent's right hand and wrap the fingers across the fleshy part of his palm below his little finger.

Twist the opponent's hand to the right while stepping in to place his hand against the chest. Apply downward pressure on the opponent's hand against the chest. Leave the opponent's hand on the chest to fully control the subject and to gain leverage.

Lean forward to use body weight to add additional pressure to the joint.

Use the left hand to further control the opponent.

Two-Handed Wristlock. Both hands can be used in the wristlock to maximize the leverage and pressure needed to bend and twist the joint. To execute the two-handed wristlock, Marines—

Place both thumbs on the back of the opponent's hand, thumbs crossed.

Hook the fingers of both hands around the fleshy part of the opponent's palm on both sides of his hand.

Step into the opponent and apply pressure downward on the back of his hand to bend the joint and rotate his wrist away from the body to twist the joint.

Enhanced Pain Compliance on Wristlock. Enhanced pain compliance techniques are applied in the third and fourth levels in the continuum of force. Additional pain be applied to a wristlock by—

Adding downward pressure to the elbow with the other hand or elbow by using the fingers to pull in on the opponent's radial nerve located on the inside of the forearm. When pressure is added to the opponent's radial nerve, his direction can be controlled.

Applying pressure against the opponent's finger joint to bend it in a direction it cannot bend (i.e., splitting the fingers).

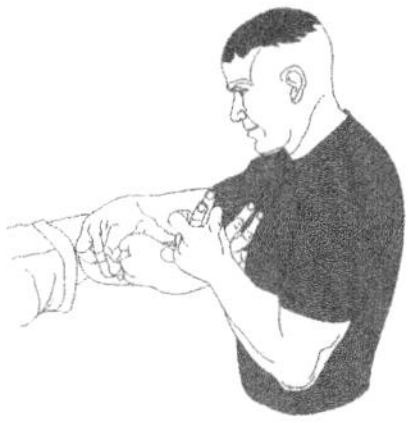

Come-Along Holds

Marines use a come-along hold to control and move an opponent.

Escort Position. A common come-along hold is the escort position. To execute the escort position, Marines—

Face the opponent. Use the left foot to step forward at a 45-degree angle. Turn to face the right side of the opponent.

Use the right hand to firmly grasp the opponent's right wrist. With the left hand, firmly grasp the opponent's right triceps.

Position the opponent's controlled arm diagonally across the torso, keeping his wrist against the right hip. The Marine should be standing to the right of and behind the opponent.

Note: This technique works well when escorting an opponent on either the right or left side. Take caution when escorting an opponent by ensuring his controlled hand is not in a position to grab the holstered weapon. The preferred escort position is from the Marine's left side, so that the opponent is kept further away from the weapon.

Wristlock Come-Along. To execute the wristlock come-along, Marines—

Use the left hand to execute a basic wristlock. Incorporate the right hand in a two-handed wristlock for more control.

Maintain pressure on the opponent's wrist with the right hand, step forward, and pivot around to stand next to the opponent.

Release the left hand, quickly reach under the opponent's arm from behind, and grab his hand.

Use the left hand and apply downward pressure on the opponent's wrist.

Controlling Technique. The following controlling technique is used when an opponent grabs the Marine's wrist. To execute the technique, Marines—

Trap the opponent's hand with the palm of the other hand.

Rotate the opponent's trapped hand up and on his forearm while maintaining downward pressure on his trapped hand.

Apply downward pressure with both hands until the opponent is taken to the ground.

Armbars

An armbar is a joint manipulation in which pressure is applied on a locked elbow, just above the joint, in the direction the joint will not bend. An armbar has to be locked in quickly, but still requires a slow, steady pressure to gain compliance.

Basic Armbar. To execute a basic armbar, Marines—

Use the right hand to grab the opponent's right wrist.

Bring the left hand down on or above the opponent's elbow joint. To gain additional leverage, pivot to face the opponent.

Use the left hand to apply downward pressure on or above the opponent's elbow joint while pulling up on his wrist.

Armbar from a Wristlock. To execute an armbar from a wristlock, Marines—

Use the right hand to grab the opponent's right hand and execute a reverse wristlock.

Bring the left hand down on or above the opponent's elbow joint.

Use the left hand to apply downward pressure on or above the opponent's elbow joint while pulling up on his wrist.

Takedowns

A takedown is used to bring an opponent to the ground to further control him.

Takedown From a Wristlock Come-Along. To take the opponent to the ground from a wristlock come-along, Marines—

Use the right foot to push down on the opponent's calf or Achilles tendon.

Maintain control of the opponent's wrist and elbow and apply a slow, steady pressure to bring him to the ground.

Armbar to a Takedown. This technique is used to take a noncompliant opponent to the ground from an armbar. To execute the armbar takedown, Marines—

Use the right hand to execute a reverse wristlock.

Bring the left hand or forearm down on or above the opponent's elbow joint.

Pivot so the back is facing the opponent and, at the same time, lift the left elbow and slide the body so it is against the opponent, placing the armpit high above the opponent's elbow joint.

Lean back, placing the body weight on the opponent's arm until he complies or is taken to the ground.

Note: This technique may break the opponent's arm. Therefore, this technique should not be employed if the objective is a nonlethal takedown.

Wristlock Takedown. This technique is used to take a noncompliant opponent to the ground from a basic wristlock and to put him in a position where he can be handcuffed, if necessary. To execute the takedown, Marines—

Use the right hand to execute a basic wristlock. Incorporate the left hand in a two-handed wristlock.

Apply downward pressure on the wristlock, pivot on the ball of the right foot, and quickly turn to the right to take the opponent to the ground.

Continue to apply pressure on the wrist joint as the opponent lands on his back with his arm straight in the air.

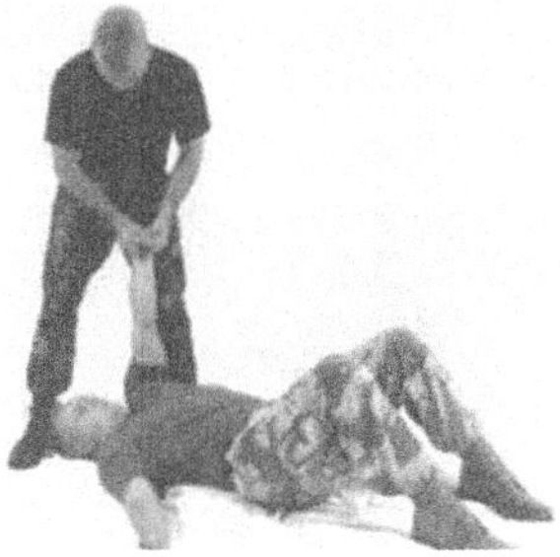

Slide the left foot under the opponent's back, underneath his armpit.

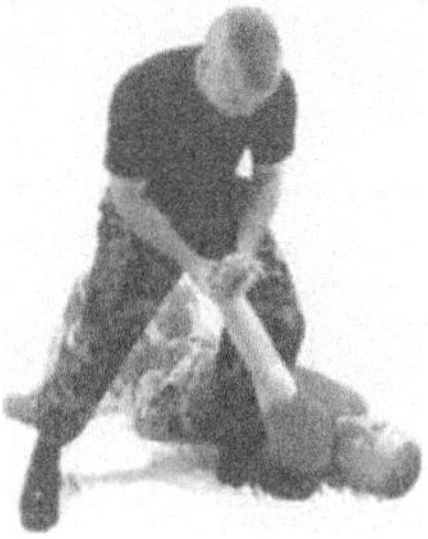

Apply pressure with the knee against the opponent's triceps while pulling back on his arm.

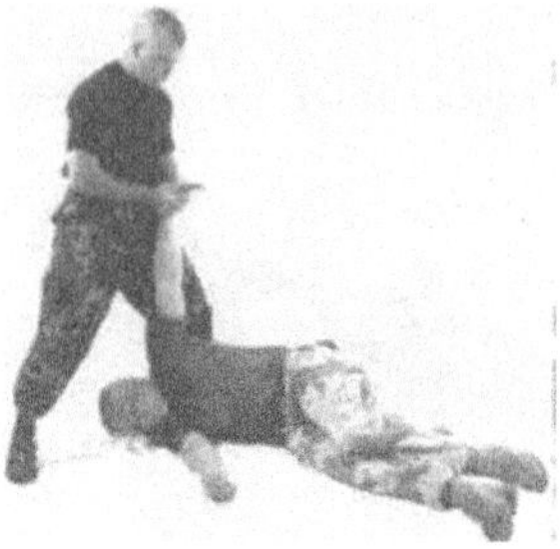

Continue applying pressure with the knee on the opponent's arm. Pivot and step around the opponent's arm to roll him on his stomach.

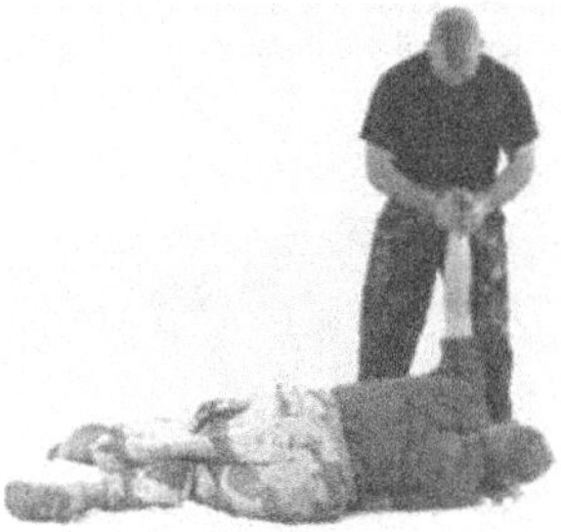

Kneel down with one knee on the opponent's back. The other knee is placed on the opponent's neck and shoulder on either side of his arm. Apply inward pressure with the knees to lock his arm in place.

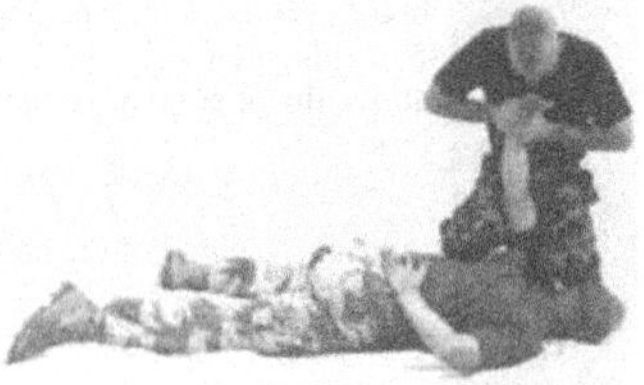

Tell the opponent to put his other hand in the middle of his back. Bring the opponent's controlled hand to the center of his back.

Escort Position Takedown. This technique can be used to control a noncompliant opponent from the escort position. To execute the escort position takedown, Marines—

Lock the opponent's arm straight across the body while rotating his wrist away from the body.

Use the left hand or forearm and apply downward pressure above the opponent's elbow where the triceps meet.

Step back with the right foot and, keeping the opponent's hand controlled against the hip, pivot to the right while continuing to apply downward pressure on his arm to bring him to the ground.

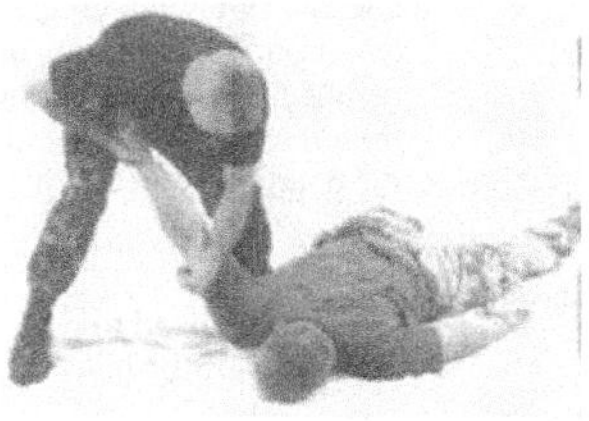

Note: This technique works well when escorting a subject on the Marine's right or left side. When taking down a subject from the right side, step back and pivot to the left.

2. Nonlethal Baton

A baton or nightstick can be an effective compliance tool when used correctly. Batons can be used defensively (blocking), offensively (striking), and as a restraining device when needed. In the fourth level of the continuum of force (assaultive [bodily harm]), defensive tactics include baton or nightstick blocks and blows. Blows to the head or other bony parts of the body are considered deadly force. When deadly force is not authorized, Marines must be able to employ blocks, strikes, and restraints with the baton with the minimum of force.

Grips

One-Handed Grip. To execute the one-handed grip, Marines—

Use the right hand to grasp the lower end of the baton, about 2 inches from the end.

Wrap the thumb and index finger around the baton so they are touching one another. The grip on the baton should be firm, but natural.

Two-Handed Grip. To execute the two-handed grip, Marines—

Use the right hand to grasp the lower end of the baton, about 2 inches from the end. Wrap the thumb and index finger around the baton so they touch one another.

Use the left hand to grasp the upper end of the baton, palm down, about 2 inches from the end. The hands should be approximately 10 to 12 inches apart.

Stance and Method of Carry

The basic warrior stance serves as the foundation for initiating nonlethal baton techniques. The method of carry provides effective defensive positions with a wide range of options to control a combative opponent.

One-Handed Carry. To execute the one-handed carry, Marines—

Grip the baton using the one-handed grip.

Elevate the baton, with the gripping hand at a level between the belt and shoulder.

Keep the left hand in the position of the basic warrior stance.

Two-Handed Carry. This carry is effective for blocks. To execute the two-handed carry, Marines—

Grip the baton using the two-handed grip.

Elevate the baton, with the left hand higher than the right hand.

Orient the weapon toward the subject.

Movement

In a nonlethal confrontation, movement may be made to create distance between Marines and the opponent or to close the gap to control him. When facing an opponent, Marines move in a 45-degree angle to either side of the opponent. Moving at a 45-degree angle is the best way to both avoid an opponent's strike and to put Marines in the best position to control the opponent.

Blocking Techniques

One-handed blocks are used when carrying the baton in the one-handed carry. The same one-handed blocks used in combative stick techniques (see page 3-6) apply in nonlethal baton engagements. The blocks include blocks for a vertical strike, a forward strike, and a reverse strike. Because the baton is often carried with two hands, there are also two-handed blocks that are effectively used from this carry. Two-handed blocks are discussed in the following subparagraphs.

High Block. Marines execute a high block to deter a downward vertical attack directed at the head and shoulders. To execute the high block, Marines—

Raise the baton up to a level even with or above the head. The baton should be in a horizontal position to block the blow. The fingers of the left hand should be open and behind the baton.

Place the baton perpendicular to the opponent's striking surface to absorb the impact of the blow.

Bend the elbows slightly to help absorb the impact of the blow. The arms should give with the strike of the blow.

Low Block. Marines execute a low block to deter an upward vertical attack directed at the abdomen, groin, or torso. The opponent's blow can be delivered by a foot, knee, or fist. To execute the low block, Marines—

Lower the baton to a level even with or below the groin. The baton should be in a horizontal position to block the blow. The fingers of the left hand should be open and behind the baton.

Place the baton perpendicular to the opponent's striking surface to absorb the impact of the blow.

Bend the elbows slightly to help absorb the impact of the blow. The arms should give with the strike of the blow.

Right Block. Marines execute a right block to deter an attack directed at the head, neck, flank, or hip. The opponent's blow can be delivered by a foot, knee, fist, or elbow. To execute the right block, Marines—

Thrust the baton in a vertical position to the right side.

Pivot to the right by stepping with the left foot and pivoting off the ball of the right foot. Rotate the hips and shoulders into the direction of the block. The fingers of the left hand should be open and behind the baton.

Place the baton perpendicular to the opponent's striking surface to absorb the impact of the blow.

Bend the elbows slightly to help absorb the impact of the blow. The arms should give with the strike of the blow.

Left Block. Marines execute a left block to deter an attack directed at the head, neck, flank, or hip. The opponent's blow can be delivered by a foot, knee, fist, or elbow. To execute the left block, Marines—

Thrust the baton in a vertical position to the left side.

Pivot to the left by stepping with the right foot and pivoting off the ball of the left foot. Rotate the hips and shoulders into the direction of the block. The fingers of the left hand should be open and behind the baton.

Place the baton perpendicular to the opponent's striking surface to absorb the impact of the blow.

Bend the elbows slightly to help absorb the impact of the blow. The arms should give with the strike of the blow.

Middle Block. Marines execute either a left or right block to deter an attack directed at the face, throat, chest, or abdomen. To execute the middle block, Marines—

Thrust the baton in a vertical position straight out in front of the body. The fingers of the left hand should be open and behind the baton.

Place the baton perpendicular to the opponent's striking surface to absorb the impact of the blow. The baton should be held with the left hand forward of the right.

Bend the elbows slightly to help absorb the impact of the blow. The arms should give with the strike of the blow.

Restraining Technique

The strong-side armlock is used to restrain an opponent who is not compliant. To execute the strong-side armlock, Marines—

Use the right hand to run the baton up under the opponent's left armpit, parallel to the ground.

Use the right foot to step forward at a 45-degree angle to the left side of the opponent. The baton is across his forearm.

Use the right hand to drive the baton forward and up so the action bends the opponent's arm behind his back. At the same time, continue moving around him to get behind him.

Place the baton on the opponent's forearm with the thumb and/or fingers and apply pressure to his forearm. At the same time, grasp the other end of the baton with the left hand.

Pull up on the low end of the baton with the right hand. At the same time, push down on the top end of the baton with the left forearm, reaching around with the left hand to grasp the opponent's biceps or shoulder.

Continue exerting downward pressure with the left forearm while pulling back on the opponent's biceps with the left hand. This places the opponent in a position where he is controlled and can be moved.

Apply pressure with the foot against the bend in the opponent's leg above his calf. This lowers the opponent to the ground rather than throwing him to the ground and risking severe injury.

Striking Target Areas

Marines must avoid striking an opponent in the head, neck, or other bony parts with the baton because this is considered deadly force and can lead to serious bodily injury or death. Instead, the legs, arms, and buttocks are target areas that are considered nonlethal.

Legs. Primary targets are the thighs and lower legs. Avoid striking the knee and ankle joints because this can cause permanent damage.

Arms. Primary targets are the upper arms. Avoid striking the shoulder and elbow and wrist joints because this can cause permanent damage.

Buttocks. Primary targets are the buttocks. Avoid striking any other part of the torso, including the chest, rib cage, spine, tail bone, and groin because strikes to these areas can cause permanent damage or death.

One-Handed Striking Techniques

One-Handed Forward Strike. A forward strike follows either a horizontal line or a downward diagonal line using a forehand stroke. To execute the one-handed forward strike, Marines—

Stand facing the opponent with the baton carried in a one-handed carry.

Place the right hand palm up, swing the baton from right to left, and make contact with the opponent.

One-Handed Reverse Strike. A reverse strike follows either a horizontal line or a downward diagonal line using a backhand stroke. To execute the one-handed reverse strike, Marines—

Stand facing the opponent with the baton carried in a one-handed carry.

Bend the right arm and cross the arm to the left side of the body. The baton should be close to or over the left shoulder.

Place the right hand palm down, swing the baton from left to right, and make contact with the opponent.

Two-Handed Striking Techniques

Two-Handed Forward Strike. This strike is an effective follow-up to a middle block or left block. To execute the two-handed forward strike, Marines—

Pull back with the left hand while driving the right hand forward toward the opponent. The baton should be horizontal to the ground. Power is generated by stepping forward with the right foot and rotating the right hip and shoulder into the strike.

Contact the opponent with the end of the baton.

Two-Handed Reverse Strike. This strike is an effective follow-up to a middle block or left block. To execute the two-handed reverse strike, Marines—

Pull back with the right hand while driving the left hand forward toward the opponent. The baton should be horizontal to the ground. Power is generated by stepping forward slightly with the left foot and rotating the left hip and shoulder into the strike.

Contact the subject with the end of the baton.

Front Jab. This strike is effective for countering a frontal attack. It can also be executed as a quick poke to keep a subject away. To execute the front jab, Marines thrust both hands forward in a quick jab. The baton is held either horizontal to the ground or at a slight downward angle.

Rear Jab. This strike is effective for countering a bear hug from the rear. To execute the rear jab, Marines thrust both hands rearward in a quick jab. The baton is held either horizontal to the ground or at a slight downward angle.

APPENDIX A

PUGIL STICK TRAINING

A pugil stick is a training device used to simulate a rifle bayonet so that effective, but safe, training can be conducted to build proficiency of rifle bayonet techniques. Pugil stick training builds on the techniques used to throw punches. Pugil stick training is the only "full contact" training provided to Marines in the Close Combat Program. Pugil stick training teaches Marines to function when faced with stress and violence, and it prepares them to deliver a blow and take a blow. It also provides them with the physical and mental skills vital to success on the battlefield.

1. Pugil Stick Training

Design

A pugil stick consists of a stick wrapped in padding at both ends that can be gripped like a rifle. The pugil stick is approximately the same weight and length of an unloaded rifle with a bayonet attached.

Holding the Pugil Stick

The pugil stick is held in the same manner as the service rifle. All movements come from the basic warrior stance. To hold the pugil stick correctly, Marines—

- Use the right hand to grasp the lower end of the pugil stick overhanded.
- Use the left hand to grasp the upper end of the pugil stick underhanded.
- Use the right forearm to lock the lower end of the pugil stick against the hip.
- Orient the blade end of the pugil stick toward the opponent.

Safety Equipment

The following safety equipment must be worn during any pugil stick bout.

Groin Protection. Groin protection protects the groin from an accidental blow. It should be pulled high around the waist to protect the groin area, with the concave portion against the body.

Flak Jacket. The flak jacket provides protection to the body. It is worn completely fastened.

Neck Roll. The neck roll prevents whiplash if Marines receive a blow to the head. The neck roll further supports the head and protects the neck from blows. The neck roll is worn above the flak jacket and below the helmet. The tied end of the neck roll faces the front.

Helmet. A regulation football helmet protects the face and head. The helmet must fit snugly and the chin straps must be adjusted and snapped.

Mouthpiece. The mouthpiece is worn on the upper teeth to protect the teeth.

Gloves. Marines wear gloves to protect the hands if pugil sticks do not have gloves built-in.

Pugil Stick Screening

Prior to pugil stick training, instructors must ask participating Marines the following questions. If a Marine answers "yes" to any question, he must see the corpsman or a competent medical authority who evaluates him and determines whether or not the he can participate in the training.

- Are you on light duty?
- Are you restricted to running shoes by a corpsman or a doctor?
- Have you fought in a pugil stick bout within the last 7 days?
- Have you received a blow to the head within the last 7 days?
- Have you had a concussion within the last 6 months?
- Have you had dental surgery within the last 24 hours?
- Do you have stitches or staples on your body?
- Have you had a shoulder or head injury within the last 5 years?
- Are you taking a prescription drug?
- Do you have an ear infection or current sinus infection?
- Have you had a broken bone within the last 6 months?

Safety Personnel

The following safety personnel are required to conduct pugil stick training:

- One close combat instructor must officiate the bout.
- One close combat instructor trainer, commissioned officer, or staff noncommissioned officer must be in the training area to serve as range safety officer.
- One corpsman must be in the training area.

Training Area

To prevent injury, Marines train on areas with soft footing (i.e., sand or grass). Training mats are not recommended because feet can stick to the mats, prohibiting movement or causing joint injuries by twisting a knee. Bouts should not take place on a hard surface area; e.g., a flight deck or parking lot. A boxing ring may be used to conduct pugil stick bouts; ring dimensions can vary as long as there is ample room to execute the techniques in the training area.

Second Impact Syndrome

Second Impact Syndrome occurs when a second blow to the head produces a second concussion that occurs within 1 week following a previous concussion (before recovery from the first concussion). Second Impact Syndrome causes rapid brain swelling and can cause death. Therefore, there must be 7 days between pugil stick bouts to reduce the risk of severe injury resulting from Second Impact Syndrome. The 7 day separation between pugil stick bouts significantly reduces the possibility of injury, particularly in someone who may have suffered a brain injury or concussion but shows no symptoms.

Any Marine who experiences headaches or the following symptoms after training must be examined by appropriate medical personnel:

- Blurred vision.
- Ringing in the ears.
- Dilation of the pupils.
- Slurred speech.
- Bleeding from ears or mouth.
- Swelling in head or neck area.
- Any unnatural discoloration in head or neck area.

The Marine should not be allowed to participate in pugil stick training or any other activity where a heavy blow might be sustained for a minimum of 7 days after the headache or other symptoms have subsided.

Safety Measures

The following safety measures must be followed:

- Gear must be worn properly throughout training.
- Contact lenses or glasses will not be worn.
- False teeth will be removed from the mouth.
- Nothing will be worn around the neck except the neck roll.

- Competition among groups of Marines is authorized as long as it does not overshadow training objectives or compromise safety procedures.
- Safety and proper techniques are paramount.
- Safety is more important than competition.

2. General Rules and Regulations Governing Pugil Stick Bouts

Instructors and Support Personnel Requirements

At a minimum, one close combat instructor observes a pugil stick bout. For safety purposes, it is better to have two instructors judging a bout because each instructor can fully observe each of the fighters. The best position for observation is to the right of a fighter. This allows the instructor to see the fighter's facial expression and body movement. The instructor's position must not interfere with the fight.

In addition, one commissioned or staff noncommissioned officer (to serve as range safety officer) and one corpsman will be in the training area.

Prior to the Bout

Prior to the bout, Marines—

- Are paired according to height, weight, and gender.
- Wear the proper safety equipment.

Once Marines have properly donned the safety equipment, they wait for instructions from the close combat instructor. On the command of the close combat instructor, a pair of Marines will enter the ring. Once in the ring, the close combat instructor inspects each Marine for the proper safety equipment.

Note: There should be enough gear so that when two Marines are training, two other Marines can be donning safety equipment for the next bout.

During the Bout

The bout begins when the close combat instructor blows the whistle. All strikes are directed above the waist. Upon hearing a whistle blast, all fighting immediately ceases.

Stopping a Bout

There are two reasons for stopping a bout: delivery of a scoring blow or an unsafe condition. The close combat instructor trainer, close combat instructor, or the range safety officer may stop the bout at any time an unsafe condition is observed.

Scoring Blow. A scoring blow is an offensive technique delivered to a vulnerable area of an opponent with sufficient force and precision to be considered as a disabling or killing blow. Scoring blows are not judged solely on the degree of force with which a blow is delivered, but on the accuracy and techniques employed. A scoring blow is defined as—

- A straight thrust with the blade end of the weapon (red end of the pugil stick) to the opponent's face mask or throat.
- A slash to the side of the opponent's helmet (below the ear) or neck with the red end of the pugil stick.
- A heavy blow to the opponent's head with an authorized technique (i.e., buttstroke, smash) using the butt (black end) of the pugil stick.

When a scoring blow is delivered, the close combat instructor blows the whistle to stop the bout.

Unsafe Condition. The bout will be stopped as soon as an unsafe condition exists. An unsafe condition exists when a Marine is unable to defend himself, loses his balance and falls down on one or two knees or falls down completely, shows instability (e.g., buckling at the knees), loses muscular tension in his neck and his head snaps back or to one side, or appears disoriented. An unsafe condition also exists if a Marine lets go of one end of the pugil stick, the equipment (e.g., helmet, neck roll) falls off, or a Marine fails to use the proper techniques.

If any of these conditions occur—

- The close combat instructor trainer, close combat instructor, range safety officer, or anyone supervising training stops the bout and separates the two Marines.
- The corpsman evaluates the possibly injured Marine to determine if the fight can continue. The corpsman—
  - Ensures that he is alert and responsive, both verbally and physically.
  - Talks to him to see if he is coherent.
  - Makes certain he comprehends and replies to verbal questions.
  - Makes certain his speech is not slurred.
  - Checks his physical signs.
  - Ensures his eyes are focused and not dazed or glazed.
  - Ensures his legs are not wobbly or shaky.

The pugil stick is not used as a baseball bat. The use of unauthorized techniques will result in expulsion from the bout. Only techniques taught in bayonet training are authorized. These techniques include—

- Straight thrust.
- Buttstroke (horizontal and vertical).
- Smash.
- Slash.
- Parry.
- Blocks (low, high, left, right).

After the Bout

A third whistle blast is used prior to identifying the Marine who delivered the killing blow.

A fourth whistle blast is used prior to demonstrating the killing technique used by the Marine who won the bout.

3. Directions for Making Pugil Sticks

Close combat instructor trainers are taught to make pugil sticks to standard specifications at school. It is recommended that all pugil sticks be made under the supervision of a close combat instructor trainer. Until the Marine Corps acquires a standardized pugil stick that units can order through the supply system, instructors use the following information and instructions to make pugil sticks locally.

Supplies

The following supplies are needed to make one pugil stick:

- One 45" by 2" circular, oak, wood dowel.
- Four 9" wide by 27" long foam pad pieces to make two inner handguards of double thickness.
- Four 11" wide by 19" long foam pad pieces to make two outer handguards of double thickness.
- Canvas material to cover inner and outer handguards.
- One foam pad cut 1/4" thick by 7" wide by 13" long to make the centerguard.
- One 8" wide by 14"long piece of canvas material to make a pugil stick sock.
- Canvas material to make two pugil stick end caps. Ideally, two different color fabrics should be used to simulate the bayonet and buttstock ends of a rifle. Common colors for end caps are black and red. Each end cap is a different color to help the officiating certified close combat instructor determine the scoring blow during the pugil stick bout.
- Two 1/4" thick by 11" wide by 54" long foam pad pieces to make two ends.
- Two foam pads cut 1/4" thick by 3" wide by 14 1/2" long to make two end plug inserts.
- One can of 3M Photo Mount™ spray adhesive (NSN 8040-01-120-4009)to adhere foam padding to the stick. (Spray adhesive is available from self-service.)
- One roll of duct tape or riggers tape (NSN 7510-00-266-5016) to protect foam pad and

canvas from wear and to reinforce gluing action of spray adhesive. (Duct tape or riggers tape is available from self-service.)

- One roll para-cord or 5-50 cord to tie both the inner and outer handguards together.
- One roll of thread strong enough to sew canvas material together.
- Sixteen grommets, one for each of the four corners of the inner and outer handguards.
- One foam sleeping mat (NSN 8465-01-109-3369). Use unserviceable foam sleeping mats obtained from Defense Reutilization and Marketing Office.

Tools

The following tools are needed to make a pugil stick:

- Tape measure, ruler, or yardstick.
- Scissors.
- Grommet machine.
- Knife.
- Sewing machine.

Directions

To make a pugil stick, Marines perform the following steps.

Preparing Materials

Step 1. Cut foam sleeping mat into the dimensions listed above. Cut mats to double thickness.

Step 2. Use spray adhesive to adhere the two 9" by 27" pieces of foam together. Use the spray adhesive to adhere the two 11" by 19" pieces of foam together. Let adhesive dry before using the double-thick foam for handguards.

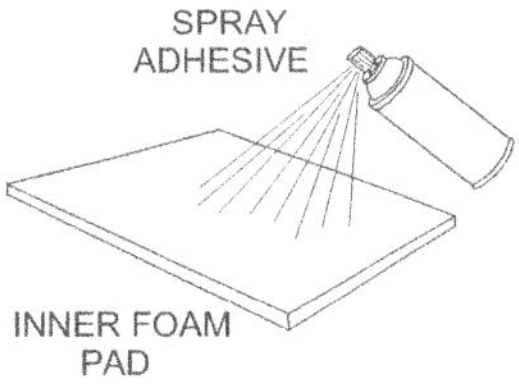

Step 3. Cut canvas material to make two coverings for the inner handguards. Cut material to make two coverings for the outer handguards. Two inner handguards and two outer handguards are required for each pugil stick.

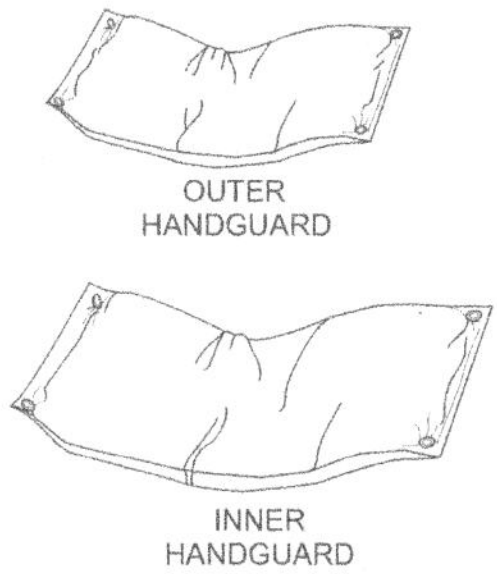

Step 4. Use a sewing machine to make a sleeve out of the 9" by 27" canvas. Sew down three sides. Insert foam pad inside the canvas sleeve and then sew down the fourth side. Use the grommet machine to create holes reinforced with metal grommets in each of the four corners. Once finished, the inner handguards are complete.

Note: It is recommended that the Fabric Repair Shop be used to sew canvas sleeves and apply grommets.

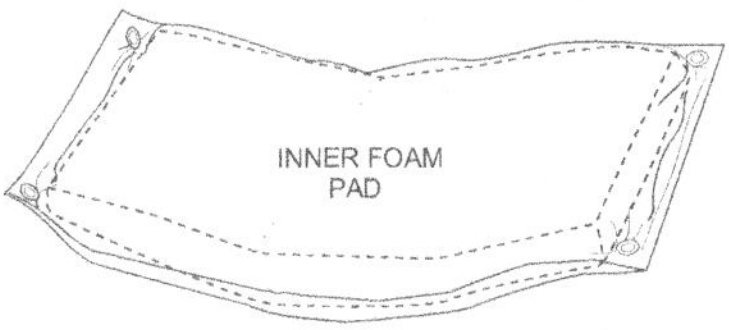

Step 5. Repeat step 4 to make the other inner handguard and both outer handguards.

Constructing Ends, End Plugs, and End Caps

Step 1. Apply spray adhesive on approximately 8 inches of one end of the dowel. This adhesive serves to glue the foam pad to the wooden dowel.

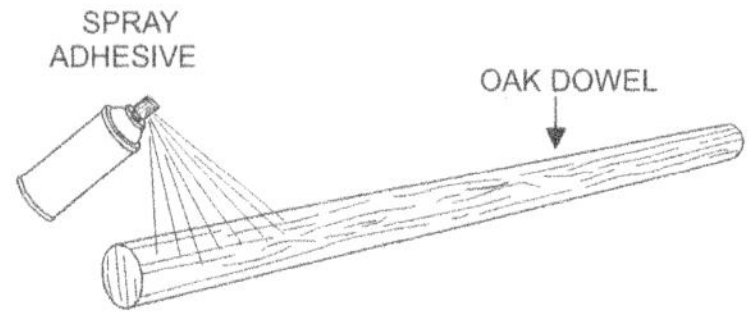

Step 2. Apply spray adhesive on one side of the 11" by 54" foam pad. This causes the foam pad to stick to itself when it is rolled around the end of the wooden dowel. Ensure 3 inches of the rolled foam pad extends past the wooden dowel for stability.

Step 3. Apply spray adhesive on one side of 3" by 14 1/2" foam pad. Roll pad into a tight cylinder.

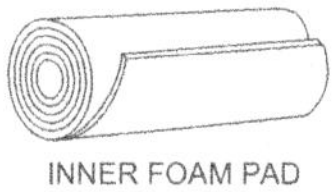

Step 4. Apply spray adhesive on the exposed end of the dowel and fill the hole in the end with the end plug. The end plug prevents the wooden dowel from protruding beyond the edge of the rolled foam padding. This is an important safety precaution to prevent injury.

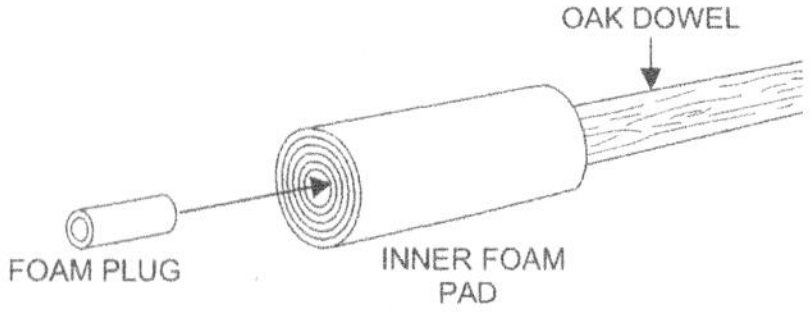

Step 5. Use duct tape or riggers tape to secure and reinforce the foam pad once the end plug is inserted in the end.

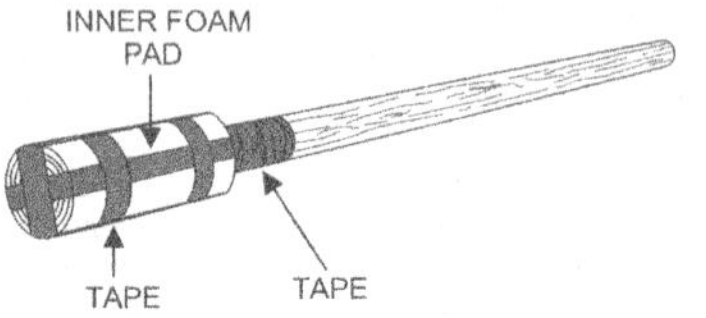

Step 6. Measure the dimensions of the ends (typically, the length of the cylinder should be 12 inches long to allow for taping). Cut and sew canvas material in the shape of a cylinder with one open end. Once complete, this becomes the end cap. The end cap is a colored sleeve that covers the rolled and taped foam pad at each end of the pugil stick.

Note: It is recommended that the Fabric Repair Shop be used to make the end cap.

Step 7. Slide one of the end caps over the rolled foam pad and tape it to the dowel.

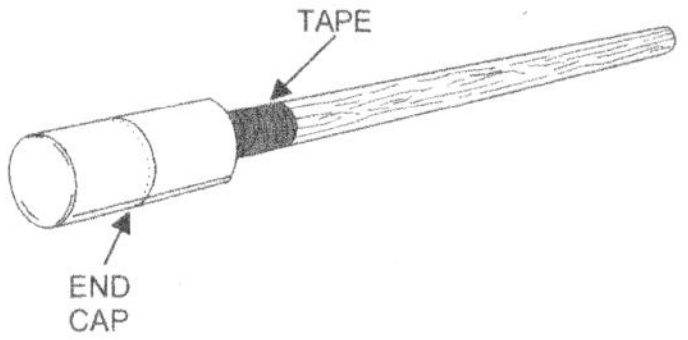

Building the Centerguard

Step 1. Spray one side of the 7" by 13" foam pad with spray adhesive and wrap it tightly around the center of the dowel.

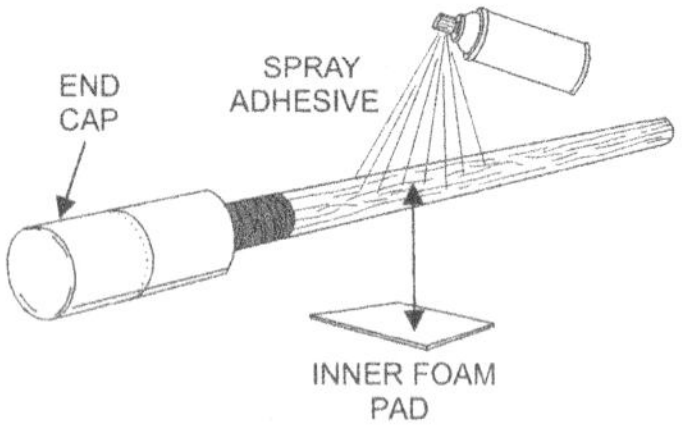

Step 2. Wrap duct tape or riggers tape around the foam pad to secure.

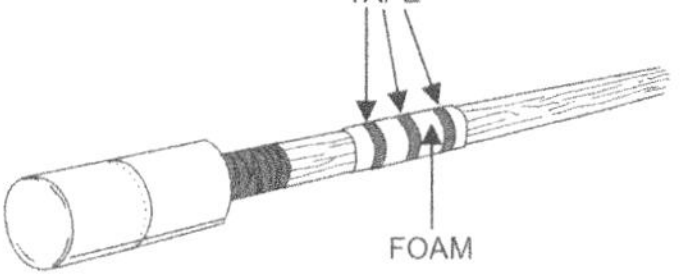

Step 3. Use a sewing machine to make a tubular pugil stick sock out of the 8" by 14" piece of canvas. Slide the pugil stick sock over the rolled foam pad and secure it with the duct tape or riggers tape. Duct tape or riggers tape protects the edges of the centerguard and reinforces the gluing effects of the spray adhesive.

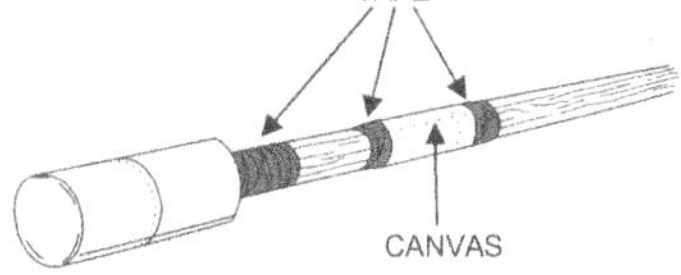

Note: It is recommended that the Fabric Repair Shop be used to make the pugil stick sock.

Finishing the Opposite End Cap

After one end cap is complete and the centerguard is finished, then repeat steps 1 through 7 in Constructing Ends, End Plugs, and End Caps for the other end of the pugil stick. Ensure that the opposite color end cap is used.

Fastening Handguards

Step 1. Place the inner handguards (longer piece) around the dowel. The inner handguard is placed on first because it protects the wrist. One inner handguard is needed for the left hand and one for the right hand.

Step 2. Fold the inner handguard around the pugil stick and run the para-cord or 5-50 cord through the metal grommets.

Step 3. Tighten down the corner ends of the inner handguards by pulling on the para-cord or 5-50 cord.

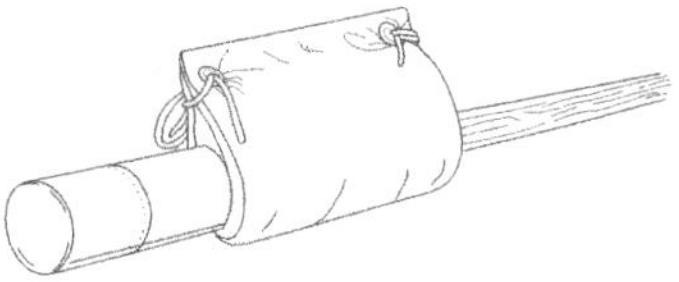

Step 4. Tie off the cord with a square knot above the metal grommets.

Step 5. Place the outer handguards (shorter piece) over the inner handguards (longer piece). Repeat steps 1-4 for the other handguard. The outer handguards provide additional protection to the knuckles and fingers and provides additional stability for the inner handguards.

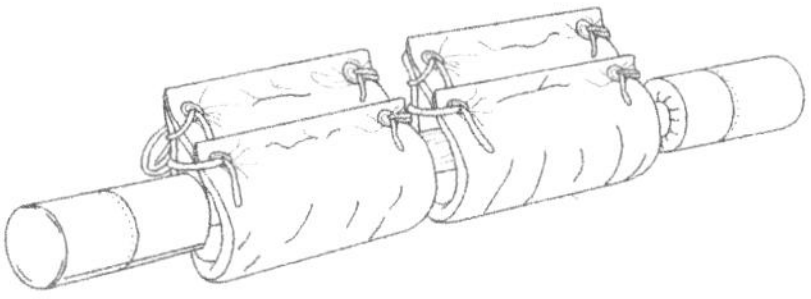

Serviceability Inspection

Prior to any training, all pugil sticks are inspected to ensure they are fit for use. If any of the following conditions exist, the pugil stick is unsafe and will not be used:

- Stick protrudes through the end caps.
- Frayed or worn edges exist on canvas surfaces.
- The stick can be felt through the pads or the padding is too soft.
- End caps or handguards are not securely attached.

(reverse blank)

APPENDIX B

SAFETY PRECAUTIONS DURING TRAINING

1. General Safety Precautions

When training close combat techniques, certain safety precautions must be adhered to to prevent injuries.

Most training should be conducted on a training area with soft footing such as a sandy or grassy area. If training mats are available, they should be used. A hard surface area is not appropriate for close combat training.

All techniques should be executed slowly at first. Marines can increase the speed of execution as they become more proficient. Marine-on-Marine training that requires contact (chokes, throws, ground fighting, and unarmed restraints and manipulation) should not be executed at full force or full speed.

If a technique is applied to the point that a Marine is uncomfortable, the Marine must "tap out." This indicates immediate release of the pressure being applied or to immediately stop the technique. The Marine "taps out" by firmly tapping his hand several times on any part of the opponent's body that will get his attention or by saying *stop*.

Second Impact Syndrome occurs when a second concussion develops within hours, days, or weeks following a prior concussion (and before recovery from the first concussion). Second Impact Syndrome causes rapid brain swelling and can cause death. Marines who experience headaches or other symptoms following training must be examined by appropriate medical personnel. These symptoms can include, but are not limited to, blurred vision, ringing in the ears, dialation of the pupils, bleeding from the ears or mouth, slurred speech, swelling in head or neck area, or any unnatural discoloration of head or neck. They should not be allowed to participate in pugil stick training or any other activity where a heavy blow might be sustained for a minimum of 7 days after the symptoms have subsided.

2. Safety Precautions for Individual Techniques

Falls

When training Marines to fall, they should progress from the ground, to a kneeling or squatting position, and then to a standing position. This ensures they are comfortable and understand the technique before progressing to executing falls from a higher profile. This instructional technique vastly reduces the risk of injury.

Strikes and Punches

When training Marines to strike and punch, they begin by executing the techniques "in the air." As they become more proficient, they execute strikes on equipment (when available) such as an air shield or a heavy bag. At no time should they be permitted to execute strikes on another student.

Chokes and Ground Fighting Chokes

When training Marines to execute chokes, they will not apply pressure to the opponent's throat during training because the trachea and windpipe can be crushed. During training, Marines should practice the proper procedures for blood chokes. They should never execute air chokes. At no time during training should a choke be applied at full force or full speed. No choke should be held for longer than 5 seconds.

Unarmed Restraints and Manipulation

When training Marines to execute unarmed restraints and manipulation techniques, they utilize slow and steady pressure. Never apply these techniques at full force or at full speed.

Pugil Stick Training

A blow to the head during training will bruise the brain. A second blow to the head can cause death. This is known as the Second Impact Syndrome. There must be 7 days between pugil stick training to prevent injury or death.

Headquarters
Department of the Army
Washington, DC
10 June 1987

Army Regulation 600–9

Effective 10 June 1987

Personnel—General

The Army Weight Control Program

By Order of the Secretary of the Army:

JOHN A. WICKHAM, JR.
General, United States Army
Chief of Staff

Official:

R. L. DILWORTH
Brigadier General, United States Army
The Adjutant General

History. The original form of this regulation was published on 1 September 1986. Since that time, changes have been issued to amend the original. As of 3 August 1987, permanent changes 1 and 2 remained in effect. This UPDATE printing incorporates all of those changes directly into the body of text. This publication has been reorganized to make it compatible with the Army electronic publishing database. No content has been changed.

Summary. This regulation implements guidance in DOD directive 1308.1, dated 29 June 1981, which establishes a weight control program in all the Services.

Applicability. This regulation applies to all members of the Active Army, the Army National Guard (ARNG), and the U.S. Army Reserve (USAR) to include those ARNG and USAR personnel in Active Guard/Reserve (AGR) status.

Proponent and exception authority. The proponent agency of this regulation is the Office of the Deputy Chief of Staff for Personnel.

Army management control process. This regulation is subject to the requirements of AR 11–2. It contains internal control provisions but does not contain checklists for conducting internal control reviews. These checklists will be contained in a DA Circular in the 11 series.

Supplementation. Supplementation of this regulation is prohibited without prior approval from HQDA (DAPE–MPA), WASH DC 20310.

Interim changes. Interim changes to this regulation are not official unless they are authenticated by The Adjutant General. Users will destroy interim changes on their expiration dates unless sooner superseded or rescinded.

Suggested Improvements. Users are invited to send comments and suggested improvements on DA Form 2028 (Recommended Changes to Publications and Blank Forms) directly to HQDA (DAPE–MPA), WASH DC 20310.

Distribution. Active Army, ARNG, and USAR: A.

Contents (Listed by paragraph and page number)

UNCLASSIFIED

Contents—Continued

Appendixes

Table List

Figure List

Contents—Continued

Section I
Introduction

1. Purpose
This regulation establishes policies and procedures for the implementation of the Army Weight Control Program.

2. References
Required publications are listed in appendix A.

3. Explanation of abbreviations and terms
Abbreviations and special terms used in this regulation are explained in the glossary.

4. Objectives
a. The primary objective of the Army Weight Control Program is to insure that all personnel—
(1) Are able to meet the physical demands of their duties under combat conditions.
(2) Present a trim military appearance at all times.
b. Excessive body fat—
(1) Connotes a lack of personal discipline.
(2) Detracts from military appearance.
(3) May indicate a poor state of health, physical fitness, or stamina.
c. Objectives of the Army Weight Control Program are to—
(1) Assist in establishing and maintaining—
(a) Discipline.
(b) Operational readiness.
(c) Optimal physical fitness.
(d) Health.
(e) Effectiveness of Army personnel through proper weight control.
(2) Establish appropriate body fat standards.
(3) Provide procedures for which personnel are counseled to assist in meeting the standards prescribed in this regulation.
(4) Foster high standards of professional military appearance expected of all personnel.

Section II
Responsibilities

5. General
The Army traditionally has fostered a military appearance which is neat and trim. Further, an essential function of day-to-day effectiveness and combat readiness of the Army is that all personnel are healthy and physically fit. Self-discipline to maintain proper weight distribution and high standards of appearance are essential to every individual in the Army.

6. Deputy Chief of Staff for Personnel (DCSPER)
The DCSPER is responsible for the Army Weight Control Program.

7. The Surgeon General (TSG)
TSG will—
a. Establish medical examination and counseling policies in support of the Army Weight Control Program.
b. Evaluate the medical aspects of the program.
c. Establish and review procedures for determination of body fat content.
d. Provide guidance on improving the nutritional status of soldiers.

8. Deputy Chief of Staff for Logistics (DCSLOG)
The DCSLOG will—
a. Establish food service guidance in support of the Army Weight Control Program.
b. Publish guidance and information pertaining to the caloric content of items served on master menus.

9. Chief, National Guard Bureau (CNGB)
The CNGB will—
a. Implement and monitor the Army Weight Control Program in the ARNG (including units).

b. Take appropriate action under guidance prescribed in this regulation.

10. Chief, Army Reserve (CAR)

The CAR will—

a. Monitor the Army Weight Control Program in the USAR.

b. Take appropriate action under guidance prescribed in this regulation.

11. Commanders of major Army commands (MACOMs)

MACOM commanders will insure that soldiers within their commands are evaluated under the body fat standards prescribed in this regulation.

12. Commanding General, US Army Training and Doctrine Command (CG, TRADOC)

The CG, TRADOC will inform personnel at initial entry on active duty (AD) about basic nutrition and sound food consumption practices.

13. Commanding General, US Army Forces Command (CG, FORSCOM)

The CG, FORSCOM will implement and maintain the Army Weight Control Program in USAR troop program units (TPUs).

14. Commanders of major medical commands

a. Commanding General, US Army Health Services Command (CG, HSC). The CG, HSC will—

(1) Institute weight reduction and counseling programs in Army medical facilities in support of the Army Weight Control Program.

(2) Provide appropriate literature and training aids for use by soldiers, supervisors, and commanders in selection of a proper diet.

b. Commanders of other major medical commands (overseas) will institute weight reduction and counseling programs in Army medical facilities in support of the Army Weight Control Program overseas.

15. Commanding General, US Army Reserve Components Personnel Center (CG, ARPERCEN)

The CG, ARPERCEN will—

a. Monitor the Army Weight Control Program in the Individual Ready Reserve (IRR).

b. Take appropriate action under guidance prescribed in this regulation.

c. Insure that members applying for tours of AD, active duty for training (ADT), active duty support (ADS), and AGR meet the body fat standards prescribed in this regulation. (Soldiers who do not meet these standards will not be permitted to enter on AD, ADT, or ADS, or in AGR status.)

16. Individuals

Each soldier (Commissioned Officer, Warrant Officer and Enlisted) is responsible for meeting the standards presecribed in this regulation. To assist soldiers in meeting these responsibilities, screening tables will continue to be prescribed for use as is currently being done. A five percent zone below the screening table weight ceiling is suggested as a help to soldiers in targeting their personal weight at a level which will minimize the probability of exceeding the screening table weight ceiling as a matter of habit. Soldiers should be coached to select their personal weight goal within or below the five percent zone and strive to maintain that weight through adjustment of life style and fitness routines. If a soldier consistently exceeds the personal weight goal, he or she should seek the assistance of master fitness trainers for advice in proper exercise and fitness; and health care personnel for a proper dietary program. In other words, exceeding a properly selected goal should "trigger" the soldier to use the substantial help available to alter the fitness and dietary behavior before confronting the finality of the screening table and initiation of official action if the body fat standards are exceeded.

17. Commanders and superviors

Commanders and supervisors (Active and Reserve Components) will—

a. Implement the Army Weight Control Program, to include evaluation of the weight and military appearance of all soldiers under their jurisdiction; to include measuring body fat as prescribed in this regulation.

b. Insure the continued evaluation of all soldiers under their command or supervision against the body fat standards prescribed in this regulation.

c. Maintain data on soldiers in their command or under their supervision who—

(1) Enter a weight control program each year.

(2) Subsequently either meet the body fat standards prescribed in this regulation or were separated from the Service for reasons related to overweight conditions.

d. Encourage soldiers to establish a personal weight goal as described above.

e. ensure that personnel responsible for issuing TDY and PCS orders include in the text of all orders the words: "You are responsible for reporting to your next duty station/school in satisfactory physical condition, able to pass the APFT and meet weight standards."

18. Health Care Personnel

Health care personnel will—

a. Assist commanders and supervisors by providing weight reduction counseling to individuals who are overweight.

b. Identify those individuals who have a pathological condition requiring medical treatment.

c. Evaluate overweight soldiers, when—

(1) A soldier has a medical limitation (DA Form 3349, AR 40–501 prescribes assignment limitations for soldiers with profiles. For example, no mandatory strenuous physical activity); or

(2) Is pregnant; or

(3) When an evaluation is requested by a unit commander (this is an option for unit commanders and is not mandatory); or

(4) Separation is being considered for failure to make satisfactory progress in a weight control program; or

(5) Six months prior to ETS.

19. Master Fitness Trainers

Master Fitness Trainers will—

a. Prescribe proper exercise and fitness techniques to assist soldiers in determining, achieving and maintaining an appropriate personal weight goal.

b. Assist commanders in developing proactive programs which clearly establish physical fitness as a unit value.

Section III
Proper Weight Control

20. Policy

a. Commanders and supervisors will monitor all members of their command (officers, warrant officers, and enlisted personnel) to insure that they maintain proper weight, body composition (as explained in the glossary), and personal appearance. At minimum, personnel will be weighed when they take the Army Physical Fitness Test (APFT) or at least every 6 months. Personnel exceeding the screening table weight (shown in table 1) or identified by the commander or supervisor for a special evaluation will have a determination made of percent body fat. Identification and counseling of overweight personnel are required.

b. Commanders and supervisors will provide educational and other motivational programs to encourage personnel to attain and maintain proper weight (body fat) standards. Such programs will include—

(1) Nutrition education sessions conducted by qualified health care personnel. These sessions are required for all soldiers enrolled in a Weight Control Program.

(2) Exercise programs, even though minimum APFT standards are achieved.

c. Maximum allowable percent body fat standards are as follows in table 2. However, all personnel are encouraged to achieve the more stringent Department of Defense (DOD)-wide goal, which is 20 percent body fat for males and 26 percent body fat for females.

Table 2
Maximum allowable percent body fat standards

Age Group	Male (% body fat)	Female (% body fat)
17–20	%20	%28
21–27	%22	%30
28–39	%24	%32
40 & Older	%26	%34

d. Personnel who are overweight (as explained in the glossary) (including soldiers who become pregnant while on the weight control program)—

(1) Will be considered nonpromotable (to the extent such nonpromotion is permitted by law).

(2) Will not be authorized to attend professional military or civilian schooling. Personnel arriving at professional military schools overweight will be denied enrollment, except:

(a) for personnel in a TDY and return or a TDY enroute to PCS status who, in the opinion of the appropriate school commander or commandant, can potentially meet their body fat standard by the end of the course or within 30 days, whichever comes first. Overweight personnel who do not meet their body fat standard within the probationary period will be disenrolled and will not graduate. An academic efficiency report as required, will be annotated to indicate failure to meet body fat standards, and the appropriate career division will be notified for those personnel in a TDY enroute to PCS status.

(b) for personnel in a PCS status, and then, only for a 30 day probationary period. If at the end of the probationary period, body fat standards have not been met, the student will be disenrolled. An academic efficiency report as required, will be annotated to indicate failure to meet body fat standards, and the appropriate career division will be notified for assignment instructions.

(3) Will not be assigned to command positions.

21. Procedures

a. Body fat composition will be determined for personnel—

(1) Whose body weight exceeds the screening table weight in table 1.

(2) When the unit commander or supervisor determines that the individual's appearance suggests that body fat is excessive.

b. Routine weigh-ins will be accomplished at the unit level. Percent body fat measurements will be accomplished by company or similar level commanders (or their designee) in accordance with standard methods prescribed in Appendix B to this regulation. Soldiers will be measured by individuals of the same gender. If this cannot be accomplished, a female soldier will be present when males measure females. IRR members on AT, ADT, and SADT will have a weigh-in, and body fat evaluation (if required) by the unit to which attached. Active and Reserve component soldiers exceeding the body fat standards in paragraph 20*c*, above, will be provided weight reduction counseling by health care personnel, entered in a Weight Control Program by unit commanders, and flagged under the provisions of AR 600–31 by the unit commander. Enrollment in a Weight Control Program starts on the day that the soldier is informed by the unit commander that he/she has been entered in a weight control program. The weight reduction counseling can be accomplished prior to or shortly after entry into a program.

c. The sample correspondence shown in figure 1 will be completed and retained by the unit commander or supervisor to document properly recommendations and actions taken in each case.

d. A medical evaluation will be accomplished by health care personnel when the soldier has a medical limitation, or is pregnant, or when requested by the unit commander. One is also required for soldiers being considered for separation due to failure to make satisfactory progress in a weight control program, or within 6 months of ETS. Aircraft crew members, who exceed the body fat standards, will be referred to a flight surgeon for possible impact on flight status. If an individual's condition is diagnosed by medical authorities to result from an underlying or associated disease process, health care personnel will take one of the following actions:

(1) Prescribe treatment to alleviate the condition and return personnel to their unit.

(2) Hospitalize individuals for necessary treatment; this action applies to Active Army personnel only. Reserve Component personnel will be referred to their personal physician for further evaluation or treatment at the individual's expense.

(3) Determine whether the individual's condition is medically disqualifying for continued service. In these cases, disposition will be made under provisions of appropriate regulations.

e. If health care personnel discover no underlying or associated disease process as the cause of the condition (or RC personnel decline to be evaluated under paragraph 21*d*) and the individual is classified as overweight, these facts will be documented and the individual entered in a weight control program by the unit commander. Suspension of favorable personnel actions will be initiated under AR 600–31 for personnel in a weight control program.

(1) The required weight loss goal of 3 to 8 pounds per month is considered a safely attainable goal to enable soldiers to lose excess body fat and meet the body fat standards described in para 20*c*. Weight-ins will be made by unit personnel monthly (or during unit assemblies for ARNG and USAR personnel) to measure progress. A body fat evaluation may also be done by unit personnel to assist in measuring progress.

(2) As an exception to *g* below, an individual who has not made satisfactory progress after any two consecutive monthly weigh-ins may be referred by the commander or supervisor to health care personnel for evaluation or reevaluation. If health care personnel are unable to determine a medical reason for lack of weight loss—and if the individual is not in compliance with the body fat standards at paragraph 20*c* and still exceeds the screening table weight (table 1)—the commander or supervisor will inform the individual that—

(a) Progress is unsatisfactory.

(*b*) He or she is subject to separation as specified in *j* below.

f. Commanders and supervisors will remove individuals administratively from a weight control program as soon as the body fat standard is achieved. The screening table weight will not be used to remove soldiers from a Weight Control Program. The removal action will be documented as shown in figure 1; removal of suspension of favorable personnel actions will be accomplished at that time.

g. After a period of dieting and/or exercise for 6 months, soldiers who have not made satisfactory progress (as explained in the glossary) and who still exceed the screening table and body fat standards will be processed as follows:

(1) If health care personnel determine that the condition is due to an underlying or associated disease process, action described in *d* above will be taken.

(2) If no underlying or associated disease process is found to cause the overweight condition, the individual will be subject to separation from the Service under appropriate regulations indicated in *j* below.

h. Personnel will be continued in a weight control program (as provided in *e* through *g* above) after the initial 6-month period if they—

(1) Still exceed the body fat standard.

(2) Have made satisfactory progress toward their weight loss (as indicated in the glossary), or are at or below the screening table weight (table 1).

(3) *For RC personnel only.* If the individual has not obtained an evaluation from his/her personal physician under the provisions of 21*d* above, and cannot demonstrate that the overweight condition results from an underlying or associated disease process, the individual may be separated under appropriate regulations without further medical evaluation by health care personnel.

i. To assist commanders and supervisors, a flow chart outlining procedural guidance is shown at figure 2.

j. The commander or supervisor will inform the individual in writing that initiation of separation proceedings is being considered under the following regulations: AR 635–200, chapter 5–15, AR 635–100, chapter 5; NGR 600–200, chapter 7; NGR 600–101; NGR 600–5; NGR 635–100; AR 135–175; or AR 135–178. This procedure will be followed unless a medical reason is found to preclude the loss of weight or there is other good cause to justify additional time in the weight control program.

(1) The individual will immediately respond to the separation consideration letter in writing. The commander or supervisor will consider the response and initiate separation action if no adequate explanation is provided, unless the individual submits an application for retirement, if eligible. USAR personnelwho fall under the purview of this paragraph will be released from AD or discharged, as appropriate.

(2) If separation action is not initiated or does not result in separation, the individual will be entered or continued in a weight control program, as specified in *e* above.

k. Following removal from a weight control program, if it is determined (under *a* above) that an individual exceeds the screening table weight (table 1) and the body fat standard prescribed in paragraph 20*c* within 36 months, the following will apply:

(1) If the unit commander determines that the individual exceeds the screening table weight and the body fat standard—

(*a*) Within 12 months from the date of the previous removal from the program and no underlying or associated disease process is found as the cause of the condition, the individual will be subject to separation from the Service under *j* above.

(*b*) After the 12th month, but within 36 months from the date of the previous removal from the program, and no underlying or associated disease process is found as the cause of the condition, the individual will be allowed 90 days to meet the standards. Personnel who meet the body fat standard after that period will be removed from the program. All others will be subject to separation from the Service under *j* above.

(*c*) Personnel who meet the AR 600–9 standards and become pregnant will be exempt from the standards for the duration of the pregnancy plus the period of convalescent leave after termination of pregnancy. They will be entered in a weight control program, if required, after completion of convalescent leave and approval of a medical doctor that they are fit for participation in a weight control program. This procedure also applies to individuals in a medical holding unit who have been hospitalized for long periods. Soldiers entered/reentered in a weight control program after pregnancy, prolonged treatment, or hospitalization will be considered to be in a new weight control program. Paragraph 21*k* of this regulation will not apply at that time.

(2) If the individual is determined to exceed the body fat standard and the condition is due to an underlying or associated disease process, action described in *d* above will be taken.

l. Inherent in the responsibility of selection boards is the obligation to select only those individuals who are considered to be physically fit to perform the duties required of them at all times. Compliance with the Army Weight Control Program as prescribed in this regulation will be considered in the selection process for promotion, professional military or civilian schooling, or assignment to command positions. Procedures for commanders and supervisors to provide current information for use by selection boards indicating whether individuals meet the prescribed standards will be included in DA regulations or issued by separate correspondence.

m. Records will be maintained in the MPRJ for personnel in weight control programs. On transfer from one unit to another, the losing commander or supervisor will forward a statement to the gaining unit with information indicating the status of the individual's participation in a weight control program. When the transfer is a permanent change of station, the unit commander's statement will be filed as a transfer documents in the Military Personnel Records Jacket, U.S. Army, under AR 640–10.

n. Upon removal from the weight control program, unit records on participation in a weight control program will be maintained in the MPRJ for a period of 36 months from date of removal. If the soldier is transferred to another unit prior to completion of 36 months, action will be taken in accordance with paragraph 21*m* above.

22. Reenlistment criteria

a. Personnel who exceed the screening table weight at table 1 and the body fat standard for their current age group in paragraph 20*c* will not be allowed to reenlist or extend their enlistment.

b. Exceptions to policy for Active Army personnel (including RC personnel on AD) are prescribed in this subparagraph. For soldiers who are otherwise physically fit and have performed their duties in a satisfactory manner, the commander exercising General Court Martial Convening Authority or the first general officer in the soldier's normal chain of command (whichever is in the most direct line to the soldier) may approve the following exceptions to policy:

(1) Extension of enlistment may be authorized for personnel who meet one of the following criteria:

(a) Individuals who have a temporary medical condition which precludes loss of weight. In such cases, the nature of on-going treatment will be documented; the extension will be for the minimum time necessary to correct the condition and achieve the required weight loss.

(b) Pregnant soldiers (except those soldiers who have medical conditions as listed in para 21*d*) who are otherwise fully qualified for reenlistment, including those with approved waivers, but who exceed acceptable standards prescribed in this regulation, will be extended for the minimum period which will allow birth of the child, plus 6 months. A clearance from the doctor tht the soldier is medically fit to participate in a Weight Control Program is required. Authority, which will be cited on DA Form 1695 (Oath of Extension of Enlistment), is AR 601–280, paragraph 3–3. On completion of the period of extension, the soldier will be reevaluated under paragraph 21.

(2) Exceptions to policy allowing reenlistment/extension of enlistment are authorized only in cases where—

(a) Medically documented conditions (para 21*d*) preclude attainment of required standards and

(b) Disability separation is not appropriate.

c. All requests for extension of enlistment for ARNG and USAR (TPU and IRR) personnel not on AD will be processed under NGR 600–200 or AR 140–111, chapter 3, as appropriate.

d. Requests for exceptions to policy will be forwarded through the chain of command, with the commander's personal recommendation and appropriate comment at each level. As a minimum, requests will include—

(1) The physician's evaluation.

(2) A record of progress in the weight control program.

(3) Current height and weight.

(4) Body fat content.

(5) Years of active Federal service.

(6) Other pertinent information.

e. Soldiers, who have completed a minimum of 18 years of Active Federal Service (AFS), may, if otherwise eligible, be extended for the minimum time required to complete 20 years AFS. Retirement must be accomplished not later than the last day of the month in which the soldier attains retirement eligibility. Application for retirement will be submitted at the time extension is authorized. Approval/disapproval authority is outlined in AR 601–280.

f. USAR soldiers, who have completed a minimum of 18 years of qualifying service for retired pay at age 60, may be extended for the minimum time required to complete 20 years qualifying service. Approval/disapproval authority is outlined in AR 140–111. Transfer to the IRR, Retired Reserve, or discharge will be accomplished at the end of the retirement year (RYE) in which the soldier attains the 20 qualifying years.

g. ARNG soldiers, who have completed a minimum of 18 years qualifying service for retired pay at age 60, may be extended for the minimum time required to complete 20 years qualifying service by the State TAG, disapproval authority is C, NGB. Transfer to the IRR, Retired Reserve, or discharge will be accomplished at the end of the retired year (RYE) in which the soldier attains the 20 qualifying years.

Table 1
Weight for Height Table (Screening Table Weight)

Height (in inches)	Male Age				Female Age			
	17–20	21–27	28–39	40+	17–20	21–27	28–39	40+
58	—	—	—	—	109	112	115	119
59	—	—	—	—	113	116	119	123
60	132	136	139	141	116	120	123	127
61	136	140	144	146	120	124	127	131
62	141	144	148	150	125	129	132	137
63	145	149	153	155	129	133	137	141
64	150	154	158	160	133	137	141	145
65	155	159	163	165	137	141	145	149
66	160	163	168	170	141	146	150	154
67	165	169	174	176	145	149	154	159
68	170	174	179	181	150	154	159	164
69	175	179	184	186	154	158	163	168
70	180	185	189	192	159	163	168	173
71	185	189	194	197	163	167	172	177
72	190	195	200	203	167	172	177	183
73	195	200	205	208	172	177	182	188
74	201	206	211	214	178	183	189	194
75	206	212	217	220	183	188	194	200
76	212	217	223	226	189	194	200	206
77	218	223	229	232	193	199	205	211
78	223	229	235	238	198	204	210	216
79	229	235	241	244	203	209	215	222
80	234	240	247	250	208	214	220	227

Notes:

1. The height will be measured in stocking feet (without shoes), standing on a flat surface with the chin parallel to the floor. The body should be straight but not rigid, similar to the position of attention. The measurement will be rounded to the nearest inch with the following guidelines:
   a. If the height fraction is less than 1/2 inch, round down to the nearest whole number in inches.
   b. If the height fraction is 1/2 inch or greater, round up to the next highest whole number in inches.
2. The weight should be measured and recorded to the nearest pound within the following guidelines.
   a. If the weight fraction is less than 1/2 pound, round down to the nearest pound.
   b. If the weight fraction is 1/2 pound or greater, round up to the next highest pound.
3. All measurements will be in a standard PT uniform (gym shorts and T-shirt, without shoes).
4. If the circumstances preclude weighing soldiers during the APFT, they should be weighed within 30 days of the APFT.
5. Add 6 pounds per inch for males over 80 inches and 5 pounds for females for each inch over 80 inches.

(Letterhead)

Office Symbol *Date*

SUBJECT: Weight Control Program

(Individual Soldier)

1. You have been determined to exceed the body fat standard and a goal of 3-8 pounds of weight loss per month is considered to be satisfactory progress. Failure to make satisfactory progress or achieve the body fat standards could result in separation from the service.

2. You have been flagged under the provisions of AR 600–31 and entered in a Weight Control Program.

Signature of Unit Commander

Office Symbol 1st End *Date*

FROM: Individual Soldier

TO: Commanding Officer

I understand my responsibilities to achieve the body fat standards to have my weight recorded periodically or during unit training assemblies (UTA) as applicable.

Soldier's Signature

Office Symbol 2d End

SUBJECT: Weight Control Program

FROM: Unit Commander

TO: MEDDAC

1. . . . exceeds the weight for height tables by . . . pounds and exceeds the body fat standards by .percent.

2. It is requested that a medical evaluation be conducted in view of the following (check applicable block):

( )Soldier's profile. ( )Pregnancy. ( )Unit Commander's special request. ( )Initiation of separation action (failure to make satisfactory progress in a Weight Control Program). ( )Within 6 months of ETS.

Signature of Unit Commander

Office Symbol 3d End

SUBJECT: Weight Control Program

FROM: Health Care Personnel

TO: Commanding Officer

1. In accordance with AR 600–9 . . . has been examined and found to be fit for participation in a Weight Control/Physical Exercise Program.

2. The cause of the overweight condition (is) (is not) due to a medical condition.

3. The following action(s) is/are recommended:

( )Initiation or continuation in a weight reduction program.

( )Medical treatment for pathological medical disorder (refer Reserve Component soldiers to their personal physician for evaluation at their own expense).

Signature of Health Care Personnel

Figure 1. Sample correspondence for Weight Control Program

Office Symbol 4th End

FROM: Unit Commander

TO: MEDDAC (Health Care Personnel)

1. . . . exceeds the screening table weight by . . . and body fat standards by . . . percent.

2. Nutrition education and weight reduction counseling are requested in accordance with AR 600–9 (para 20b (1)).

Signature of Unit Commander

Office Symbol 5th End

SUBJECT: Weight Control Program

FROM: Health Care Personnel

TO: Commanding Officer

1. . . . has been provided nutrition and weight reduction counseling in accordance with AR 600–9.

2. Follow-up counseling should be provided at unit level using information in Appendix C of this regulation and the assistance of master fitness trainers, if available.

Signature of Health Care Personnel

Office Symbol 6th End

FROM: Unit Commander

TO: Military Personnel Officer

1. The soldier has been determined to be in compliance with the provisions of AR 600–9, and is therefore removed from the Weight Control Program effective this date.

2. The individual's current weight is . . . pounds. Screening table weight ceiling is . . . pounds for present age category. Body fat content is . . . percent, which is within the AR 600–9 standards.

3. This correspondence will be retained in the individual's MPRJ for 36 months from this date.

Signature of Unit Commander

Figure 1. Sample correspondence for Weight Control Program—Continued

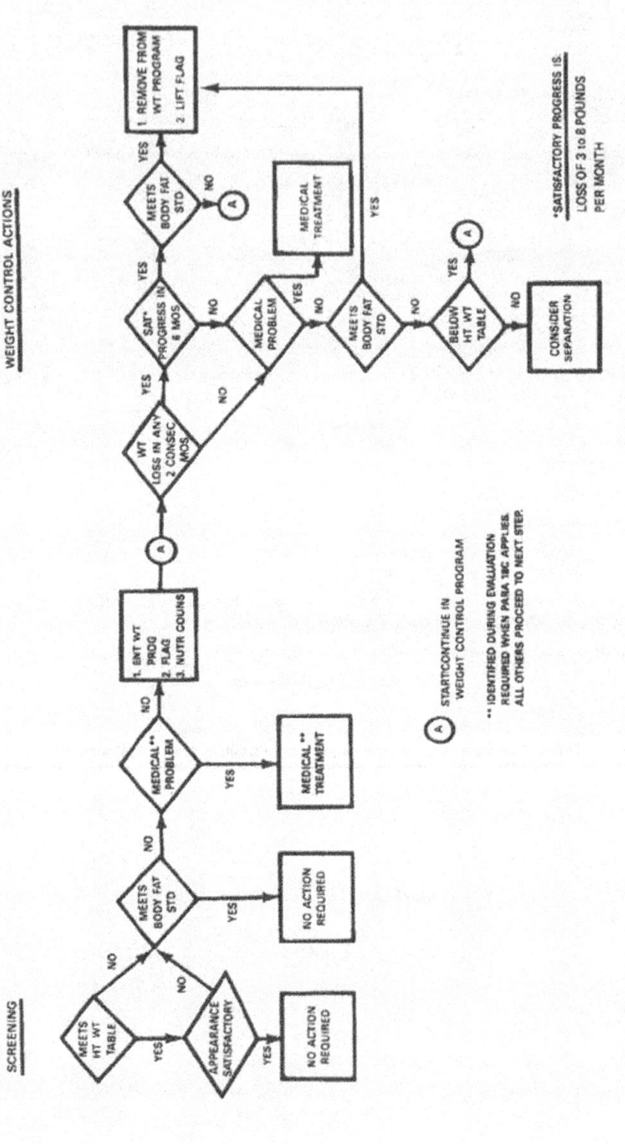

Figure 2. Flow process guide for screening and weight control actions (applies to all personnel; officers, warrant officers, enlisted)

Appendix A
References

Section I
Required Publications

AR 135–175
Separation of Officers. (Cited in paragraph 20*j*.)

AR 135–178
Separation of Enlisted Personnel. (Cited in paragraph 20*j*.)

AR 140–111
Enlistment and Reenlistment. (Cited in paragraph 21*c*.)

AR 600–31
Suspension of Favorable Personnel Actions for Military Personnel in National Security Cases and Other Investigations or Proceedings. (Cited in paragraph 20*e*.)

AR 601–280
Army Reenlistment Program. (Cited in paragraph 21.)

AR 635–100
Officer Personnel. (Cited in paragraph 20*j*.)

AR 635–200
Enlisted Personnel. (Cited in paragraph 20*j*.)

AR 640–10
Individual Military Personnel Records. (Cited in paragraph 20*m*(1).)

NGR 600–5
Enlisted Career Management. (Cited in paragraph 20*j*.)

NGR 600–101
Commissioned Officers—Federal Recognition and Related Personnel Activities. (Cited in paragraph 20*j*.)

NGR 600–200
Enlisted Personnel Management. (Cited in paragraphs 20*j* and 21*d*.)

NGR 635–100
Termination of Appointment and Withdrawal of Federal Recognition. (Cited in paragraph 20*j*.)

Section II
Related Publications
This section contains no entries.

Section III
Prescribed Forms
This section contains no entries.

Section IV
Referenced Forms
This section contains no entries.

Appendix B
Standard Methods for Determining Body Fat Using Body Circumferences, Height and Weight

B–1. Introduction

a. The procedures for the measurements of height, weight, and specific body circumferences for the estimation of body fat are described in this appendix.

b. Although circumferences may be looked upon by untrained personnel as easy measures, they can give erroneous results if proper precautions are not followed. The individual taking the measurements must have a thorough understanding of the appropriate body landmarks and measurement techniques. Unit commanders should require that designated personnel have hands-on training and read the instructions regarding technique and location, and practice before official determinations are made. Two members of the unit should be utilized in the taking of measurements, one to place the tape measure and determine measurements, the other to assure proper placement and tension of the tape, as well as to record the measurement on the worksheet. The individual taking the measurements should be of the same sex as the soldier being measured; the individual who assists the measurer and does the recording may be of either sex. The two should work with the soldier between them so the tape is clearly visible from all sides. Measurements will be made three times, in accordance with standard body measurement procedures. This is necessary for reliability purposes, since the greater number of measurements, the lesser the standard of deviation. Also, if only two measurements were taken, there would be no way to tell which measurement was the most accurate. If there is greater than 1/4-inch difference between the measurements, then continue measuring until you have three measurements within 1/4-inch of each other. An average of the scores that are within 1/4-inch of each other will be used.

c. When measuring circumferences, compression of the soft tissue is a problem that requires constant attention. The tape will be applied so that it makes contact with the skin and conforms to the body surface being measured. It should not compress the underlying soft tissues. Note, however, that in the hip circumference more firm pressure is needed to compress gym shorts. All measurements are made in the horizontal plane, (i.e., parallel to the floor), unless indicated otherwise.

d. The tape measure should be made of a non-stretchable material, preferably fiberglass; cloth or steel tapes are unacceptable. Cloth measuring tapes will stretch with usage and most steel tapes do not conform to body surfaces. The tape measure should be calibrated, i.e., compared with a yardstick or a metal ruler to ensure validity. This is done by aligning the fiberglass tape measure with the quarter inch markings on the ruler. The markings should match those on the ruler; if not, do not use that tape measure. The tape should be 1/4- to 1/2-inch wide (not exceeding 1/2-inch) and a minimum of 5–6 feet in length. A retractable fiberglass tape is the best type for measuring all areas. Tapes currently available through the Army Supply System (Federal Stock Number 8315–00–782–3520) may exceed the 1/2-inch width limits and could slightly impact on circumferential measurements. Efforts are being made to replace the supply system tape with a narrower retractable tape. In the interim, the current Army supply system or any other fiberglass tape not to exceed 5/8-inch may be used if retractable tapes cannot be purchased by unit budget funds available and approved by installation commanders.

B–2. Height and weight measurements

a. The height will be measured with the soldier, in stocking feet (without shoes) and standard PT uniform, i.e., gym shorts and T- shirt, standing on a flat surface with the head held horizontal, looking directly forward with the line of vision horizontal, and the chin parallel to the floor. The body should be straight but not rigid, similar to the position of attention. Unlike the screening table weight this measurement will be recorded to the nearest 1/4-inch in order to gather a more accurate description of the soldier's physical characteristics.

b. The weight will be measured with the soldier in a standard PT uniform, i.e., gym shorts and a T-shirt. Shoes will not be worn. The measurement should be made on scales available in units and recorded to the nearest pound with the following guidelines:

(1) If the weight fraction of the soldier is less than 1/2-pound, round down to the nearest pound.

(2) If the weight fraction of the soldier is 1/2-pound or greater, round up to the next whole pound.

B–3. Description of circumference sites, and their anatomical landmarks and technique

a. All circumference measurements will be taken three times and recorded to the nearest 1/4-inch (or 0.25). Each sequential measurement should be within 1/4-inch of the next or previous measurement. If the measurements are within 1/4-inch of each other, derive a mathematical average to the nearest quarter (1/4) of an inch. If the measurements differ by 1/4-inch or more continue measurements until you obtain three measures within 1/4-inch of each other. Then average the three closest measures.

b. Each set of measurements will be completed sequentially to discourage assumption of repeated measurement readings. For males, complete 1 set of abdomen and neck measurements, NOT three abdomen circumferences followed by three neck circumferences. Continue the process by measuring the abdomen and neck in series until you have three sets of measurements. For females, complete one set of hip, forearm, neck, and wrist measurements, NOT 3 hip followed by three forearm etc. continue the process by measuring hip, forearm, neck, and wrist series until you have 3 sets of measurements.

c. Worksheets for computing body fat are at figure B–1 (males) and figure B–3 (females). Local reproduction is authorized. A blank copy of DA Forms 5500–R and 5501–R is located at the back of this volume. These forms will be reproduced locally on 8½ x 11-inch paper. Supporting factor tables are located at tables B–1 and B–2 (males) and tables B–3 through B–8 (females) and include specific steps for preparing body fat content worksheets.

d. Illustrations of each tape measurement are at figure B–2 (males) and figure B–4 (females). A training videotape (TVT 8–103) is also available at Visual Information Libraries, and/or Training Audiovisual Support Centers (TASC).

B–4. Circumference sites and landmarks for males

a. Abdomen. The soldier being measured will be standing with arms relaxed. The abdominal measurement is taken at a level coinciding with the midpoint of the navel (belly button) with the tape placed so that it is level all the way around the soldier being measured. Record the measurement at the end of a normal expiration. It is important that the soldier does not attempt to hold his abdomen in, thus resulting in a smaller measurement. Also the tape must be kept level across the abdomen and back.

b. Neck. The soldier being measured will be standing, looking straight ahead, chin parallel to the floor. The measurement is taken by placing the tape around the neck at a level just below the larynx (Adam's apple). Do not place the tape measure over the Adam's apple. The tape will be as close to horizontal (the tape line in the front of the neck should be at the same height as the tape line in the back of the neck) as anatomically feasible. In many cases the tape will slant down toward the front of the neck. Therefore, care should be taken so as not to involve the shoulder/ neck muscles (trapezius) in the measurement. This is a possibility when a soldier has a short neck.

B–5. Circumference sites and landmarks for females

a. Neck. This procedure is the same as for males.

b. Forearm. The soldier being measured will be standing with the arm extended away from the body so that the forearm is in plain view of the measurer, with the hand palm up. The soldier should be allowed to choose which arm he/she prefers to be measured. Place the tape around the largest forearm circumference. This will be just below the elbow. To ensure that this is truly the largest circumference, since it is being visually identified, slide the tape along the forearm to find the largest circumference.

c. Wrist. The soldier being measured will stand with the arm extended away from the body so that the wrist is in plain view of the measurer. The tape will be placed around the wrist at a point above the hand just below the lower end of the bones of the forearm.

d. Hip. The soldier taking the measurement will view the person being measured from the side. Place the tape around the hips so that it passes over the greatest protrusion of the gluteal muscles (buttocks) keeping the tape in a horizontal plane (i.e., parallel to the floor). Check front to back and side to side to be sure the tape is level to the floor on all sides before the measurements are recorded. Since the soldier will be wearing gym shorts, the tape can be drawn snugly to minimize the influence of the shorts on the size of the measurement.

B–6. Preparation of the body fat content worksheets

NOTE: IT IS EXTREMELY IMPORTANT THAT YOU READ ALL OF THESE INSTRUCTIONS BEFORE ATTEMPTING TO COMPLETE THE BODY FAT CONTENT WORKSHEETS. MAKE SURE THAT YOU HAVE A COPY OF THE WORKSHEET IN FRONT OF YOU WHEN YOU ARE READING THESE INSTRUCTIONS.

a. The following paragraphs will provide information needed to perpare the Body Fat Content Worksheets for males and females, DA Form 5500–R and 5501–R, Dec 85. The worksheets are written in a stepwise fashion. The measurements and computation processes are different for males and females.

b. You will be responsible for completing a worksheet for soldiers who exceed the screening table weight (Table 1) located in this regulation, or when a unit commander or supervisor determines that the individual's appearance suggests that body fat is excessive (para 20*a* AR 600–9). The purpose of this form is to help you determine the soldier's percent body fat using the circumference technique described in this regulation.

c. Before you start, you should have a thorough understanding of the measurements to be made as outlined in this appendix. You will also need a scale for measuring body weight, a height measuring device, and a measuring tape (see specifications in para B–1*d*) for the circumference measurements.

B–7. Steps for preparing the Male Body Fat Content Worksheet, DA Form 5500–R, Dec 85

Name Print the soldier's last name, first name, and middle initial in the NAME block. Also include his Rank, and Social Security Number.

Age Print his age in years in the AGE block.

Height Measure the soldier's height as described in this appendix, to the nearest quarter of an inch, and record the measurement in the HEIGHT block.

Weight Measure the soldier's weight as described in this appendix, to the nearest pound, and record in the WEIGHT block.

Note: Follow the rules for rounding of height and weight measurements as described earlier in this appendix.

Step 1. Abdominal Measurement

Measure the soldier's abdominal circumference to the nearest quarter of an inch, and record in the block labeled "FIRST".

Step 2. Neck Measurement

Measure the soldier's neck circumference to the nearest quarter of an inch, and record in the block labeled "FIRST".

Note: REPEAT STEPS 1 and 2, in series until you have completed three sets of abdomen and neck circumferences.

Step 3. Average Abdominal Measurement

Find the mathematical average of the FIRST, SECOND, and THIRD abdominal circumferences by adding them together and dividing by three. Place this number to the nearest quarter of an inch, in the block marked AVERAGE, for STEPS 1 and 3.

Step 4. Average Neck Measurement

Find the mathematical average of your FIRST, SECOND, and THIRD neck circumferences by adding them together and dividing by three. Place this number to the nearest quarter of an inch in the block marked AVERAGE, for STEPS 2 and 4.

Step 5. Abdomen-Neck Difference

Subtract the number found in the AVERAGE block of STEP 4 from the number found in the AVERAGE block in STEP 3. Enter the result in STEP 5. This is the difference between the abdomen and neck circumferences.

Step 6. Abdomen-Neck Factor

Go to Table B–1, the Abdomen-Neck Factor Table, and locate the abdomen-neck difference in the left-most column. If the difference is a whole number, i.e., 15 inches, the Abdomen-Neck Factor is 89.93. If the difference is 15.25 inches, the factor would be 90.48, if the difference is 15.50, the factor is 91.02, and if the difference is 15.75, the factor is 91.55. Enter the appropriate factor in STEP 6.

Step 7. Height Factor

Go to Table B–2, the Height Factor Table, and locate the soldier's height in the left-most column. If the height is a whole number, i.e., 64 inches, the factor is 77.15. If the height is not a whole number, i.e., 64.25 inches, the factor is 77.27, if the height is 64.50 inches the factor is 77.39, and if the height is 64.75 inches, the factor is 77.50. Enter the appropriate factor in STEP 7.

Note: Therefore, the general rule for the factor tables is if the measure or difference is a whole number, your factor will be located under the 0.00 column, directly across from the inches column. Columns .25, .50, .75 correspond to measurements that are not whole numbers, but rather are fractions of an inch.

Step 8. Percent Body Fat

Subtract the number found in the AVERAGE block of STEP 7 from the number found in the AVERAGE block of STEP 6 and enter the difference in STEP 8. This is the soldier's PERCENT BODY FAT.

Table B–1
Male Abdomen and Neck Factor

Difference in inches	0.00	.25 (¼)	.50 (½)	.75 (¾)
5	53.44	55.06	56.61	58.09
6	59.50	60.85	62.16	63.41
7	64.62	65.78	66.91	68.00
8	69.05	70.07	71.07	72.03
9	72.96	73.87	74.76	75.62
10	76.46	77.28	78.08	78.86
11	79.63	80.37	81.10	81.82
12	82.52	83.20	83.87	84.53
13	85.17	85.81	86.43	87.04
14	87.64	88.22	88.80	89.37
15	89.93	90.48	91.02	91.55
16	92.07	92.58	93.09	93.59
17	94.08	94.57	95.05	95.52
18	95.98	96.44	96.89	97.34
19	97.78	98.21	98.64	99.06
20	99.48	99.89	100.30	100.70
21	101.10	101.49	101.88	102.26
22	102.64	103.02	103.39	103.76
23	104.12	104.48	104.83	105.19
24	105.53	105.88	106.22	106.56
25	106.89	107.22	107.55	107.87
26	108.19	108.51	108.82	109.14
27	109.44	109.75	110.05	110.35
28	110.65	110.95	111.24	111.53
29	111.82	112.10	112.39	112.67
30	112.94	113.22	113.49	113.76
31	114.03	114.30	114.56	114.83
32	115.09	115.35	115.60	115.86
33	116.11	116.36	116.61	116.85
34	117.10	117.34	117.58	117.82
35	118.06	118.30	118.53	118.77
36	119.00	119.23	119.46	119.68
37	119.91	120.13	120.35	120.57
38	120.79	121.01	121.23	121.44
39	121.66	121.87	122.08	122.29
40	122.50	122.70	122.91	123.11

Table B–2
Male Height Factor

Inches	0.00	.25 (¼)	.50 (½)	.75 (¾)
60	75.23	75.35	75.48	75.60
61	75.72	75.84	75.96	76.09
62	76.21	76.33	76.45	76.56
63	76.68	76.80	76.92	77.04
64	77.15	77.27	77.39	77.50
65	77.62	77.73	77.84	77.96
66	78.07	78.18	78.30	78.41
67	78.52	78.63	78.74	78.85
68	78.96	79.07	79.18	79.29
69	79.40	79.50	79.61	79.72
70	79.83	79.93	80.04	80.14
71	80.25	80.35	80.46	80.56
72	80.67	80.77	80.87	80.98
73	81.08	81.18	81.28	81.38
74	81.48	81.58	81.68	81.78
75	81.88	81.98	82.08	82.18
76	82.28	82.38	82.47	82.57
77	82.67	82.77	82.86	82.96
78	83.05	83.15	83.24	83.34
79	83.43	83.53	83.62	83.72
80	83.81	83.90	83.99	84.09
81	84.18	84.27	84.36	84.45
82	84.54	84.64	84.73	84.82
83	84.91	85.00	85.08	85.17
84	85.26	85.35	85.44	85.53

B–8. Steps for preparing the Female Body Fat Content Worksheet, DA Form 5501–R, Dec 85

Name Print the soldier's last name, and middle initial in the NAME block. Also include her Rank, and Social Security Number.

Age Print her age in years in the AGE block.

Height Measure the soldier's height as described in this appendix, to the nearest quarter of an inch, and record the measurement in the HEIGHT block.

Weight Measure the soldier's weight as described in this appendix, to the nearest pound, and record in the WEIGHT block.

Note: Follow the rounding rules for rounding height and weight measurements as described earlier in this appendix.

Step 1. Weight Factor

Go to Table B–3, the Weight Factor Table, and locate the soldier's weight in the left-most column, which is in 10 pound increments. If the weight is exactly 120 pounds, the factor is found under the "0" column and is 147.24. If the weight is 121 pounds, the factor is found under the "1" column and is 147.62. If the weight is 126 the factor is found under the "6" column and is 149.47. Enter the appropriate weight factor in the CALCULATIONS section, STEP 11 A.

Step 2. Height Factor

Go to Table B–4, the Height Factor Table, and locate the soldier's height in the left-most column. If the height is a whole number, i.e., 64 inches, the factor is found under the 0.00 column and is 83.75. If the height is not a whole number, i.e., 64.25 inches, the factor is 84.07. If the height is 64.50 inches, the factor is 84.40, and if the height is 64.75 inches, the factor is 84.73. Enter the appropriate height factor in the CALCULATIONS section, STEP 11 D.

Step 3. Hip Measurement

Measure the soldier's hip circumference to the nearest quarter of an inch, and record in the block labeled "FIRST."

Step 4. Forearm Measurement

Measure the soldier's forearm to the nearest quarter of an inch, and record in the block labeled "FIRST."

Step 5. Neck Measurement

Measure the soldier's neck circumference to the nearest quarter of an inch, and record in the block labeled "FIRST."

Step 6. Wrist Measurement

Measure the soldier's wrist to the nearest quarter of an inch, and record in the block labeled "FIRST."

Note: REPEAT STEPS 3, 4, 5, and 6, IN SERIES, until you have completed 3 sets of Hip, Forearm, Neck and Wrist circumferences. When you have completed this series, find the mathematical average for *each* of the 4 circumference measures and place each average in its respective AVERAGE block.

Step 7. Hip Factor

Go to Table B–5, the Hip Factor Table, and locate the soldier's AVERAGE hip circumference in the left-most column. If the circumference is a whole number, i.e., 36 inches, the Hip Factor is found in the 0.00 column and is 15.83. If the circumference is not a whole number but is 36.25 inches, the factor is 15.94. If the circumference is 36.50 the factor is 16.05. Enter the appropriate factor in the CALCULATIONS section, 11 B.

Step 8. Forearm Factor

Go to Table B–6, the Forearm Factor Table, and locate the soldier's AVERAGE forearm circumference in the left-most column. If the circumference is a whole number, i.e., 10 inches, the factor is found under 0.00 column and is 39.97. If the circumference is not a whole number but is 10.25 inches, the factor is 40.97. If the circumference is 10.75 inches, the factor is 40.97. Enter the appropriate factor in the CALCULATIONS, 11 E.

Step 9. Neck Factor

Go to Table B–7, the Neck Factor Table, and locate the soldier's AVERAGE neck circumference in the left-most column. If the circumference is a whole number, i.e., 12 inches, the factor is found under the 0.00 column and is 16.25. If the circumference is not a whole number but is 12.25 inches, the factor is 16.59. If the circumference is 12.50 inches, the factor is 16.93. If the circumference is 12.75 inches, the factor is 17.26. Enter the appropriate factor is the CALCULATIONS section, 11 F.

Step 10. Wrist Factor

Go to Table B–8, the Wrist Factor Table, and locate the soldier's AVERAGE wrist circumference in the left-most column. If the circumference is a whole number, i.e., 7 inches, the factor is found under the 0.00 column and is 3.56. If the circumference is not a whole number but is 7.25 inches, the factor is 3.69. If the circumference is 7.50 inches, the factor is 3.82. If the circumference is 7.75 inches, the factor is 3.94. Enter the appropriate factor in the CALCULATIONS section, 11 G.

Calculations

Line C. Addition of Weight and Hip Factors

Add 11 A, Weight Factor, to 11 B, Hip Factor. Enter the result on line 11 C (Total).

Line H. Addition of Height, Forearm, Neck, and Wrist Factors

Add 11 D, Height Factor, 11 E, Forearm Factor, 11 F, Neck Factor, and 11 G, Wrist Factor together. Enter the result on line 11 H. (Total).

Line I. Percent Body Fat

Subtract Line 11–H from Line–C and enter on Line I. This is the soldier's PERCENT BODY FAT.

Table B–3
Female Weight Factor

Pounds	0	1	2	3	4	5	6	7	8	9
90	134.08	134.58	135.08	135.58	136.07	136.55	137.03	137.50	137.97	138.44
100	138.90	139.35	139.80	140.25	140.69	141.13	141.56	141.99	142.42	142.84
110	143.26	143.67	144.08	144.49	144.89	145.29	145.69	146.08	146.47	146.85
120	147.24	147.62	147.99	148.37	148.74	149.10	149.47	149.83	150.19	150.54
130	150.90	151.25	151.60	151.94	152.28	152.62	152.96	153.30	153.63	153.96
140	154.29	154.61	154.94	155.26	155.58	155.89	156.21	156.52	156.83	157.14
150	157.44	157.75	158.05	158.35	158.65	158.94	159.24	159.53	159.82	160.11
160	160.40	160.68	160.96	161.25	161.53	161.80	162.08	162.35	162.63	162.90
170	163.17	163.44	163.70	163.97	164.23	164.49	164.76	165.01	165.27	165.53
180	165.78	166.04	166.29	166.54	166.79	167.04	167.28	167.53	167.77	168.02
190	168.26	168.50	168.74	168.97	169.21	169.44	169.68	169.91	170.14	170.37
200	170.60	170.83	171.06	171.28	171.51	171.73	171.96	172.18	172.40	172.62
210	172.83	173.05	173.27	173.48	173.70	173.91	174.12	174.33	174.54	174.75
220	174.99	175.20	175.41	175.61	175.82	176.02	176.22	176.42	176.62	176.82
230	177.03	177.22	177.42	177.62	177.81	178.00	178.20	178.40	178.59	178.82
240	178.97	179.16	179.35	179.54	179.73	179.92	180.10	180.29	180.47	180.66
250	180.84	181.02	181.20	181.39	181.57	181.75	181.92	182.10	182.28	182.46
260	182.63	182.81	182.98	183.16	183.33	183.51	183.68	183.85	184.02	184.19
270	184.36	184.53	184.70	184.87	185.03	185.20	185.37	185.53	185.70	185.86

Table B–4
Female Height Factor

Inches	0.00	.25 (¼)	.50 (½)	.75 (¾)
55	71.97	72.30	72.62	72.95
56	73.28	73.61	73.93	74.26
57	74.59	74.91	75.24	75.57
58	75.90	76.22	76.55	76.88
59	77.20	77.53	77.86	78.19
60	78.51	78.84	79.17	79.49
61	79.82	80.15	80.48	80.80
62	81.13	81.46	81.78	82.11
63	82.44	82.77	83.09	83.42
64	83.75	84.07	84.40	84.73
65	85.06	85.38	85.71	86.04

Table B–4
Female Height Factor—Continued

Inches	0.00	.25 (¼)	.50 (½)	.75 (¾)
66	86.36	86.69	87.02	87.35
67	87.67	88.00	88.33	88.65
68	88.98	89.31	89.64	89.96
69	90.29	90.62	90.94	91.27
70	91.60	91.93	92.25	92.58
71	92.91	93.23	93.56	93.89
72	94.22	94.54	94.87	95.20
73	95.52	95.85	96.18	96.51
74	96.83	97.16	97.49	97.81
75	98.14	98.47	98.80	99.12
76	99.45	99.78	100.10	100.43
77	100.76	101.09	101.41	101.74
78	102.07	102.39	102.72	103.05
79	103.38	103.70	104.03	104.36
80	104.68	105.01	105.34	105.67
81	105.99	106.32	106.65	106.97
82	107.30	107.63	107.96	108.28
83	108.61	108.94	109.26	109.59
84	109.92	110.24	110.57	110.90

Table B–5
Female Hip Factor

Inches	0.00	.25 (¼)	.50 (½)	.75 (¾)
30	13.19	13.30	13.41	13.52
31	13.63	13.74	13.85	13.96
32	14.07	14.18	14.29	14.40
33	14.51	14.62	14.73	14.84
34	14.95	15.06	15.17	15.28
35	15.39	15.50	15.61	15.72
36	15.83	15.94	16.05	16.15
37	16.26	16.37	16.48	16.59
38	16.70	16.81	16.92	17.03
39	17.14	17.25	17.36	17.47
40	17.58	17.69	17.80	17.91
41	18.02	18.13	18.24	18.35
42	18.46	18.57	18.68	18.79
43	18.90	19.01	19.12	19.23
44	19.34	19.45	19.56	19.67
45	19.78	19.89	20.00	20.11
46	20.22	20.33	20.44	20.55
47	20.66	20.77	20.88	20.99
48	21.10	21.21	21.32	21.43
49	21.54	21.65	21.76	21.87
50	21.98	22.09	22.20	22.31
51	22.42	22.53	22.64	22.75
52	22.86	22.97	23.08	23.19
53	23.30	23.41	23.52	23.63

Table B–6
Female Forearm Factor

Inches	0.00	.25 (¼)	.50 (½)	.75 (¾)
5	19.98	20.98	21.98	22.98
6	23.98	24.98	25.98	26.98
7	27.98	28.98	29.98	30.97
8	31.97	32.97	33.97	34.97
9	35.97	36.97	37.97	38.97
10	39.97	40.97	41.97	42.97
11	43.96	44.96	45.96	46.96
12	47.96	48.96	49.96	50.96
13	51.96	52.96	53.96	54.96
14	55.95	56.95	57.95	58.95
15	59.95	60.95	61.95	62.95

Table B–7
Female Neck Factor

Inches	0.00	.25 (¼)	.50 (½)	.75 (¾)
5	6.77	7.11	7.45	7.79
6	8.12	8.46	8.80	9.14
7	9.48	9.82	10.16	10.49
8	10.83	11.17	11.51	11.85
9	12.19	12.53	12.86	13.20
10	13.54	13.88	14.22	14.56
11	14.90	15.23	15.57	15.91
12	16.25	16.59	16.93	17.26
13	17.60	17.94	18.28	18.62
14	18.96	19.30	19.63	19.97
15	20.31	20.65	20.99	21.33

Table B–8
Female Wrist Factor

Inches	0.00	.25 (¼)	.50 (½)	.75 (¾)
5	2.54	2.67	2.80	2.93
6	3.05	3.18	3.31	3.43
7	3.56	3.69	3.82	3.94
8	4.07	4.20	4.33	4.45
9	4.58	4.71	4.83	4.96
10	5.09	5.22	5.34	5.47
11	5.60	5.72	5.85	5.98
12	6.11	6.23	6.36	6.49
13	6.62	6.74	6.87	7.00
14	7.12	7.25	7.38	7.51
15	7.63	7.76	7.89	8.01

M TAB TAB TAB TAB TAB

BODY FAT CONTENT WORKSHEET *(Male)*

For use of this form, see AR 600-9; the proponent agency is DCSPER

NAME *(Last, First, Middle Initial)*	SSN	RANK	NOTE
DOE JOHN	198-00-0000	SGT	1/4" = .25 1/2" = .50 3/4" = .75
HEIGHT *(to nearest 0.25 inch)*	WEIGHT *(to nearest pound)*	AGE	
72	175	21	

STEP	FIRST	SECOND	THIRD	AVERAGE *(to nearest 0.25 in.)*
1. Measure abdomen at the level of the navel *(belly button)* to the nearest 0.25 inch. *(Repeat 3 times.)*	36.25	36.50	36.25	36.25
2. Measure neck just below level of larynx *(Adam's apple)* to the nearest 0.25 inch. *(Repeat 3 times.)*	16.25	16.25	16.50	16.25
3. Enter the average abdominal measurement to the nearest 0.25 inch.				36.25
4. Enter the average neck measurement to the nearest 0.25 inch.				16.25
5. Subtract Step 4 from Step 3 *(Enter result)* to the nearest 0.25 inch.				20.00
6. Find result from Line 5 *(the difference between Neck and Abdomen)* in Table B-1 *(Abdomen-Neck Factor)*. Enter factor.				99.48
7. Find the height in Table B-2 *(Height Factor)*. Enter factor.				80.67
8. Subtract Step 7 from Step 6 *(Enter result)*. This is Soldier's Percent Body Fat.				18.81

REMARKS

CHECK ONE
✓ Individual is in compliance with Army Standards; ____ is not in compliance with the standards.
____ Recommended monthly weight loss is 3-8 lbs.

PREPARED BY *(Signature)*	RANK	DATE	APPROVED BY SUPERVISOR *(Printed Name and Signature)*	RANK	DATE

DA FORM 5500-R, DEC 85

Figure B–1. Sample of a completed DA Form 5500

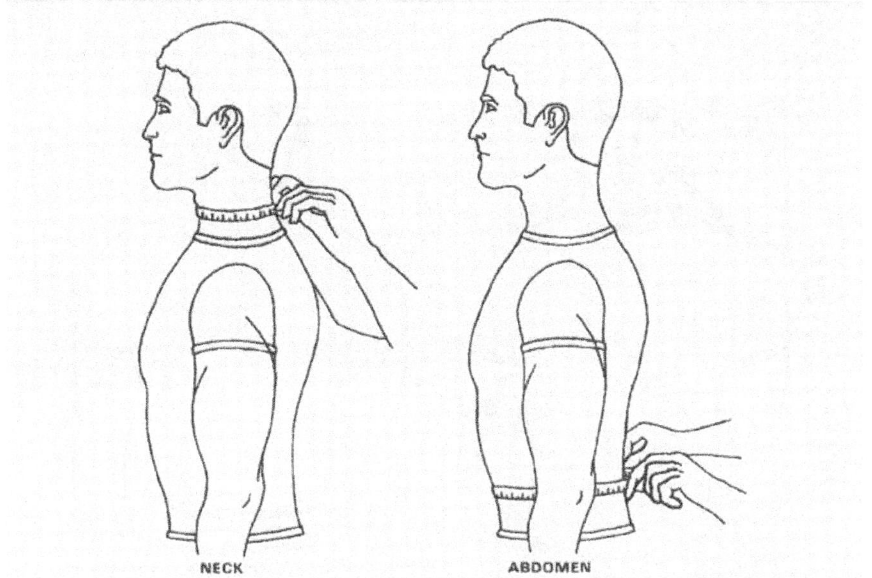

Figure B–2. Male Measurement

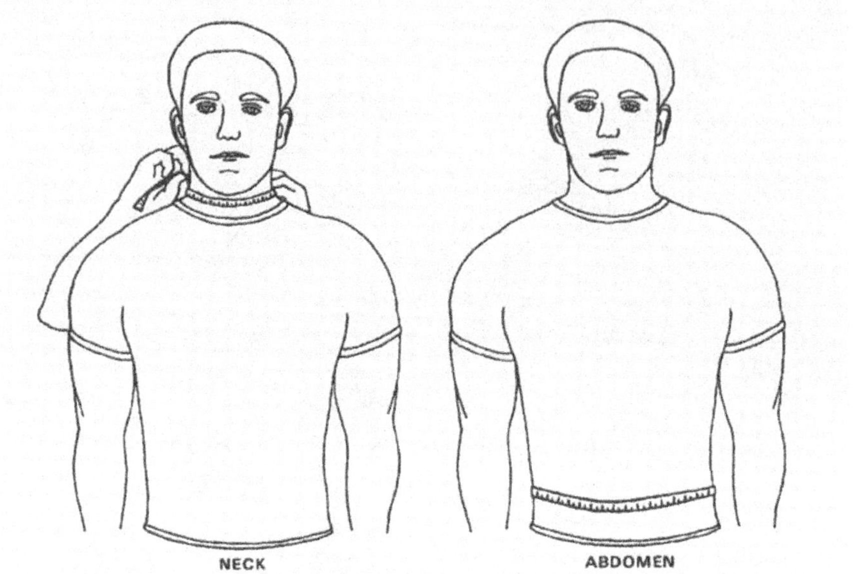

Figure B–2. Male Measurement—Continued

M TAB TAB TAB TAB

BODY FAT CONTENT WORKSHEET *(Female)*

For use of this form, see AR 600-9; the proponent agency is DCSPER

NAME *(Last, First, Middle Initial)*	SSN	RANK	NOTE:
DOE, JANE	432-00-0000	SGT	¼" = .25
HEIGHT *(to nearest 0.25 inch)*	WEIGHT *(to nearest pound)*	AGE	½" = .50
63.00	132	21	¾" = .75

STEP	FIRST	SECOND	THIRD	AVERAGE *(to nearest 0.25 in.)*
1. Find the soldier's weight in Table B-3 *(Weight Factor)*. Enter factor in 11A below				
2. Find soldier's height in Table B-4 *(Height Factor)*. Enter factor in 11D below.				
3. Measure hips at point where the gluteus muscles *(buttocks)* protrude backward the most. Round off to nearest 0.25 inch. Repeat three times, then average.	29.75	30.00	30.25	30.00
4. Measure forearm at its largest point *(with arm horizontal, palm up)* to nearest 0.25 inch. Repeat three times, then average.	8.25	8.50	8.75	8.50
5. Measure neck just below level of larynx *(Adam's apple)* to nearest 0.25 inch. Repeat three times and average.	13.25	13.50	13.75	13.50
6. Measure wrist between the bones of the hand and forearm to nearest 0.25 inch. Repeat three times, then average.	5.75	6.00	6.25	6.00
7. Find average hip measurement in Table B-5 *(Hip Factor)*. Enter factor in 11B below.				
8. Find average forearm measurement in Table B-6 *(Forearm Factor)*. Enter factor in 11E below.				
9. Find average neck measurement in Table B-7 *(Neck Factor)*. Enter factor in 11F below.				
10. Find average wrist measurement in Table B-8 *(Wrist Factor)*. Enter factor in 11G below.				

11. CALCULATIONS			REMARKS
A. Weight factor	151.60		
B. Hip factor	13.19		
C. **TOTAL** *(11A+11B)*		164.79	
D. Height factor	82.44		
E. Forearm factor	33.97		
F. Neck factor	18.28		
G. Wrist factor	3.05		
H. **TOTAL** *(11D+E+F+G)*		137.74	
I. **SOLDIER'S PERCENT BODY FAT** *(Line 11C—11H)*		27.05	

CHECK ONE

✓ Individual is in compliance with Army Standards; ____ is not in compliance with the standards.

____ Recommended monthly weight loss is 3–8 lbs.

PREPARED BY *(Signature)*	RANK	DATE	APPROVED BY SUPERVISOR *(Printed Name and Signature)*	RANK	DATE

DA FORM 5501-R, DEC 85

Figure B–3. Sample of a completed DA Form 5501

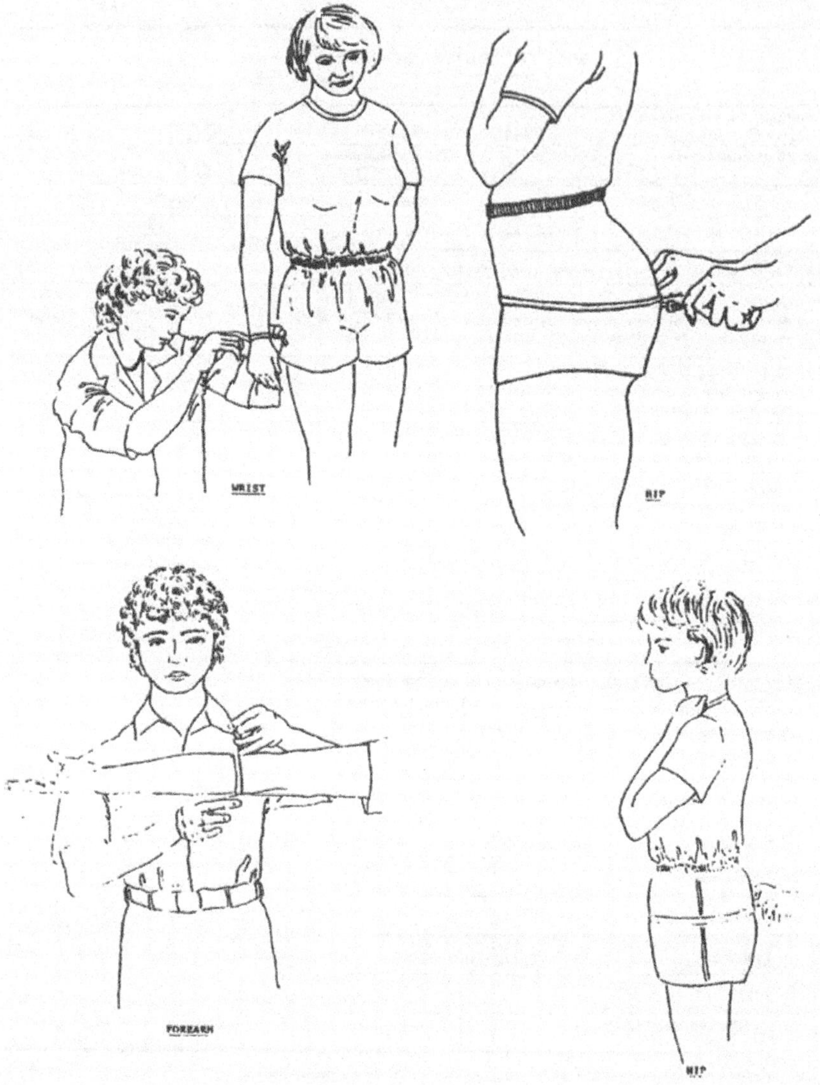

Figure B–4. Female Measurement

Appendix C
Nutrition Guide to the Weight Control Program

C–1. General
This appendix explains the basic principles of weight loss while maintaining normal nutrition. It does not replace the requirement for Active and Reserve Component Units with soldiers exceeding the body fat standards to be provided weight reduction counseling by qualified health care personnel. This guide should be used as a supplement to weight reduction counseling and as a guide for commanders in developing an effective weight control program.

C–2. Weight control—it's time to make the fitness connection
In many cases fitness begins with weight reduction. Aerobic fitness is related to an individual's body fat. The higher the fat, the less likely the individual is to be aerobically fit and the harder it is to maintain higher levels of physical stamina and endurance. A fit Army is a lean Army. Military readiness demands weight control—the spin-off is less likelihood of developing heart disease, high blood pressure, and diabetes. Proper nutrition and regular exercise are necessary to help you lose weight and improve your state of fitness.

a. Invest in yourself.

(1) Make a decision to lose weight and shape up.

(2) Get motivated.

(3) Develop a strategy (diet, exercise routine, lifestyle changes, etc.).

(4) Carry out the strategy.

b. Enjoy the payoff.

(1) A healthy appearance.

(2) An improved self image.

(3) A sense of accomplishment.

(4) A feeling of pride.

C–3. Making nutrition work
Improving your nutrition will increase your mileage in many ways. You can even lose weight while improving your performance. Your nutrition program should include the right number of calories to cause a steady loss of body fat with no loss of energy. Stay away from food fads—they're usually boring, unhealthy, too strict and lead only to temporary weight loss. Compare the benefits of a sound nutrition with the consequences of crash diets as shown below:

Table C–0–1
Nutrition

Sound Nutrition: Provides all required nutrients	**Crash Diets:** Most often lack some nutrients
Sound Nutrition: Gradual loss of body fat (1 to 2 pounds per week)	**Crash Diets:** Rapid loss of body water and muscle mass—not body fat
Sound Nutrition: Reinforces a good mental outlook	**Crash Diets:** Symptoms of grumpiness, headaches, anxiety and fatigue
Sound Nutrition: Improves health	**Crash Diets:** Can cause depression, dehydration, potential serious illness and a slow-down in your body's metabolism (the rate your body burns calories)
Sound Nutrition: Gives a sense of accomplishment	**Crash Diets:** End in eventual weight loss and failure
Sound Nutrition: Develops permanent good eating habits	**Crash Diets:** Encourage unhealthy eating habits of temporary duration

C–4. Basic strategy checklist—a plan for making the right connection

a. Good attitude. Having a good mental attitude is necessary to succeed in any program. To lose weight, a good mental attitude helps self-discipline—an important ingredient.

b. Sensible nutrition. A diet of adequate essential nutrients is necessary to prevent mental and physical fatigue. Crash diets don't work in a permanent weight control program.

c. Regular exercise. Exercise promotes physical fitness. It improves flexibility, strength, endurance and weight loss by speeding up the body's metabolism. It has also been proven to help supress the appetite.

d. Diet and exercise master plan. Create your own daily food intake and exercise plan. Keep a record. Make adjustments. You are in control.

e. Rest. Adequate rest improves attitude, posture, and appearance. Lack of rest and sleep can weaken resistance and will power.

f. A set goal. See a long range objective, then get going by setting easier-to-reach short-term goals. After the first few goals are met, the objective will be in plain view.

C–5. Balance check

If you need to lose weight, it's time to check your energy balance to see what's tilting the scale. When your energy input (calories) is greater than your energy output (activity), you store the extra calories as body fat. To stay in balance:

a. Increase exercise (frequency and intensity), or

b. Decrease calorie intake, or

c. Combine exercise with calorie reduction for best results.

C–6. Obesity risks

Excess body fat is harmful to your health. It increases your risk for developing high blood pressure, diabetes, heart disease, respiratory infections, gall bladder disease, low-back pain, and some forms of cancer. It has further drawbacks in physical appearance and interferes with physical performance.

C–7. Tailoring a nutrition program

The best nutrition program is one that allows you to lose body fat while you eat regular wholesome foods in controlled portion sizes. Does your diet include essential nutrients? Here's an easy way to check: (See below).

Table C–0–2
Diet structures

Food group	Recommended Servings Per day	Your diet
Milk (8 ozs)	2	—
Meat (3 ozs)	2	—
Bread or Cereal (1 slice or ½ cup)	4	—
Vegetables ½ cup)	2	—
Fruits (1 medium or 1/2 cup)	2	—
Does your diet provide the right amount of calories?		

Table C–0–3
Weight maintenance formula

________________	X	__________	=	________________
Your ideal body weight		Activity factor		Your calorie level for maintenance

a. Weight maintenance.

b. Activity factors.

(1) *Sedentary.* Twelve to 14 calories per pound are required if you are not involved in exercise on the job or off duty.

(2) *Active.* Fifteen calories per pound are required when your job involves physical work and/or you are engaged in a regular exercise program.

(3) *Highly active.* Sixteen to 18 calories are required per pound when very physically demanding work and/or high level of physical training is done routinely. (Most people do not fit in this category).

c. Weight loss.

(1) To lose 1 pound of fat per week, subtract 500 calories per day from your calorie maintenance level.

(2) To lose 2 pounds of fat per week, subtract 1,000 calories per day. For example:

Table C–0–4
Weight loss formula

______________	less	______________	=	______________
Calorie level for maintenance. (A above)		Calories to meet your weekly goal. (B above)		Your daily calorie level for weight loss.

C–8. What about calories

Calories don't deserve a bad name unless your intake is greater than your requirement. Calories are provided by certain nutrients in the foods and beverages you consume.

a. The following nutrients provide calories:

Table C–0–5
Nutrient-calories guide

Nutrient: Carbohydrate (Provides an efficient fuel source for the body.)
Calories: 4 per gram

Nutrient: Protein (Provides material to repair and build tissues.)
Calories: 4 per gram

Nutrient: Fat (Provides essential fatty acids and concentrated energy source for the body.)
Calories: 9 per gram

Notes:
Alcohol, while not a nutrient does provide calories (7 per gram)

b. Listed in the table below are the nutrients that do not provide calories.

Table C–0–6
Vitamins–Minerals–Water

Nutrient: Vitamins (Needed to utilize the food you eat.)
Calories: 0

Nutrient: Minerals (Needed for bones, teeth, and chemical functions.)
Calories: 0

Nutrient: Water (Necessary for life.)
Calories: 0

C–9. Good nutrition—a personal choice

The amount of calories you consume depends on the type of food you choose, its preparation and the amount you eat (PORTION SIZE). Some foods are very concentrated in calories, (i.e., the portion size is small for the amount of calories it contains). Examples are fats, candy, fried foods, most deserts, and alcohol. These foods are also of low nutrient density—they provide few nutrients for the amount of calories provided. (See below for comparison.)

Table C–0–7
High calorie-low nutrient foods

	Calories
1 piece of pecan pie	750
1 large apple	80
One-half *fried* chicken breast	400
One-half *baked* chicken breast	180
1 potato, French *fried*	275
1 *baked* potato	100
ONE-HALF CUP *syrup pack* peaches	100
One-half cup *unsweetened* peaches	50

C–10. Easy ways to save calories

One sure way to save calories without reducing PORTION SIZE is to choose foods that are:

a. Lower is sugar. (Each teaspoon supplies an additional 20 calories).

b. Lower in fat. (Each teaspoon supplies an additional 45 calories).

c. See table C–1 and table C–2 for easy ways to save calories. The Right Image of Me in the Military (TRIMM) sample meals are lower in fat and sugar—thus more "nutrient dense" and lower in calories. See table C–3 for the 1200 calorie Daily Menu and table C–4 for the 1500-calorie Daily Menu.

C–11. Portion control

a. You'll need to learn to correctly estimate portion sizes in order to ensure adequate nutrition and to control intake. A guide to estimate portion sizes is shown below.
2 ounces=1 slice meat 3 x 4 inches and 1/4-inch thick (cooked)
3 ounces=1 meat pattie or portion (1/5 of a pound)
3-inch diameter, 1/2-inch thick (cooked)
3 ounces=1/2 small chicken breast
1 ounce=1 small chicken drumstick
2 ounces=1 chicken thigh
1 ounce=1/4 cup chopped meat (tuna, spagetti meatsauce, chili, ground meat)

b. Some equal measurements are as follows:

Table C–0–8
Equal measurements

1 cup	=	8 ounces
1 tablespoon	=	3 teaspoons
1/4 cup	=	4 tablespoons
1 ounce	=	40 grams
1 liter	=	34 ounces
.7 liter	=	24 ounces
.5 liter	=	17 ounces
.2 liter	=	7 ounces

c. There is no single food that is so high in calories that a small amount cannot be eaten occasionally. Many people, however, have a particular food obsession that must be recognized. For them trying to eat "just a cookie, piece of candy, or sparerib" is too tempting. The urge to eat "the whole thing" becomes too great. You have to make and follow your own rules according to your ability to control what you eat. Avoidance is one means of control. But if you plan your diet, and diet according to your plan, you can include a favorite high calorie food item as a special occasional treat.

C–12. "Good cookin' for good lookin'"— A memo to the cook for cutting calories during food preparation

a. The milk and cheese group.

(1) Use skim or lowfat milk in recipes when making puddings, sauces, soups, and baked products.

(2) Substitute plain, unsweetened lowfat yogurt or blenderized lowfat cottage cheese in recipes that call for sour cream or mayonaise.

b. The meat, poultry, fish, and dry beans group.

(1) TRIM fat from meat. Cook meats on rack so that fat can drain off.

(2) Roast, bake, broil, or simmer meat, poultry or fish without adding fat. Braise in covered pan on stove top or pan broil in a nonstick pan; and add spices to enhance flavors.

(3) Remove skin from chicken or turkey.

(4) Chill meat broth until fat turns light and solid on top. With a spoon or knife, skim or peel fat off and discard.

c. The vegetable and fruit group.

(1) Steam, boil, broil, or bake vegetables. Some fruits may be broiled or heated with spices added for flavor.

(2) Go easy on sauces, butter, and margarine. Season with herbs and spices. Crisp-cooked vegetables usually don't require as much seasoning as over-cooked vegetables.

(3) Try lemon juice or vinegar on salads. Cut way back on regular salad dressings. (One-fourth cup creamy dressing is approximately 340 calories!)

(4) Read nutrition information labels on food packages.

d. The bread and cereal group.

(1) Use less fat and sugar than called for in recipes. Substitute lower calorie ingredients.

(2) Avoid recipes for baked products that require large amounts of fat and sugar.

(3) Check ingredient labels for fat and sugar content. Check nutrition information label for total calories in each portion.

(4) Use diet margarine or plain yogurt on baked potatoes instead of margarine, butter, or sour cream.

(5) Have boiled, steamed, or baked rather than fried potatoes.

C–13. Dining tips

a. Avoid gravies, sauces, and deep-fried food. If the meat has been fried (southern style chicken, schnitzel), remove the coating and eat only the meat.

b. Remove all the visible fat from the meat.

c. Request diet salad dressing, vinegar, or lemon juice for your salad; most restaurants have them.

d. Starchy foods are not fattening when consumed in moderate quantities. However, avoid those prepared in cream sauces or deep-fried. For example, baked potatoes with a small amount of sour cream or margarine is a good choice. Also, noodles, rice, macaroni, or spaghetti are good potato substitutes.

e. Avoid rich desserts, ice cream, jello, pastry, candy, cookies, pies, cakes, sugar, honey, jam, jelly, regular soda, and other sweets. These are sources of concentrated calories that quickly cause your total intake to skyrocket in just a few bites. USE SPARINGLY, if you must.

f. If you MUST have a snack, have fresh fruits, a few crackers or pretzels, or delicious low-calorie raw vegetables.

g. Low-calorie beverages, black coffee, unsweetened sodas, or mineral water add no calories to your diet. Lowfat milk is also a nutritious choice. Carry individually packaged sugar substitutes to sweeten beverages. Try a slice of lemon or lime in a glass of ice water.

h. Alcohol does have calories as shown below:

Table C–0–9
Alcohol calories

	Calories
American beer, 12 oz.	160
European beer, 1/2 liter	250
Cocktails, 4–6 oz. (1/2 cup)	165
Hard liquor, 1½ oz. Jigger	110
Dry wine, .25 litre (8 oz.)	200
Sweet wine, .25 liter (8 oz.)	300

C–14. Educate your appetite

a. Follow these tips to lose weight and body fat.

(1) Eat S-L-O-W-L-Y

(2) Consume less fat.

(3) Take smaller portions

(4) Consume less sugar.

(5) Take smaller bites

(6) Chew food thoroughly

(7) Eat at least three regular meals per day.

(8) Plan snacks.

b. Plan your food intake and abide by your plan. Keep a food diary and monitor your own intake.

c. Always check your intake for balanced nutrition and total calories.

d. Become aware of how many calories you're consuming- especially in snack foods. This sample calorie count shows you why:

Table C–0–10
Sample calorie chart

	Calories
Chocolate milkshake (8 oz.)	840
Coke, soda, or sugared beverage, 12 ozs.	160
French fries, 20 pieces (2″ x 1/2″)	275
Fruit pie (1/6th of 9-inch pie)	410
Pecan pie (1/6th of 9-inch pie)	750
Cheesecake (1/6th of 9-inch pie)	800
Ice cream, 2 scoops	200
Potato or corn chips, 1 ounce package	180
Chocolate candy, 1 ounce	150
Grapefruit or orange juice, 1 cup	100
Big Mac	557
Quarter-pounder with cheese	521
AAFES Jumbo Cheeseburger	654
Cottage cheese, creamed, 1 cup	223
Beer (European), 1/2 liter	250
Creamy salad dressing, 1/4 cup	340
Salter nuts:	
peanuts, 2 tbs	170
cashews, 25 nuts	252
mixed, 25 nuts	188

C–15. Some final advice

a. Get motivated. Forget those old excuses: "But I've got a large frame—you know, big bones," or "Everyone in my family is big—it's hereditary,"or "Gosh, I haven't weighed 180 pounds since I was 15 years old,"or "I can pass the APFT, so why lose weight?"

b. If you're serious about losing weight, it will be easy to give up these typical excuses. Dieting is up to you—if you are willing to try. Don't be angry that you have to lose weight; think of it as something you have chosen to do. You may be a good soldier now, but you want to strive to be the best.

c. Try not to be food centered. Eating should be a source of enjoyment and satisfaction. It shouldn't be a substitute for feelings of boredom, anger, loneliness, or discouragement. Occupy your time with other activities not related to food. This is important if you plan on being successful at losing weight. Make up your mind to control food instead of letting food control you.

d. Construct new habit patterns. Make your meals last longer, slow down, enjoy flavors, pause halfway through meals and don't stuff yourself. Make meals as pleasant as possible even if you are cutting down on what you eat.

e. Don't be discouraged by weight plateaus (i.e., periods when no weight loss occurs despite your dieting and exercise efforts). Your body is adjusting to your new eating habits and changes are taking place. Stick with it!

f. You are always responsible for what you are eating. Don't cheat; remember, overeating hurts no one but yourself.

g. If you need extra help with your diet contact the nutrition clinic at your nearest treatment facility. For an exercise plan the physical therapy clinic at your nearest medical treatment facility is an excellent resource; master fitness trainers are also excellent resources to consult for basic nutrition information and exercise program development.

h. Use the attached "Personal Weight Loss Progress Chart" DA Form 5511–R which is located in the back of this handbook to keep track of your progress.

i. Remember all the work it took to get you to your desired weight. Don't let this be wasted effort.

j. For best results, combine dietary plans with regular exercise to:

k. Your dining facility provides low calorie menu plans to assist in your weight control efforts (see tables C–3 and C–4 for sample menu patterns).

BE☆ALL☆YOU☆CAN BE—NUTRITIONALLY!

Table C–1
Weight Control a Personal Choice

Instead of Amount	Calories	Substitute with	Calories
1 cup sweetened applesauce	232	unsweetened applesauce	100
3 oz. beef bologna	237	lean ham	103
1 oz. natural Swiss or cheddar cheese	110	part-skim milk mozzarella	90
1 cup cream-style cottage cheese	223	low-fat (1%) cottage cheese	125
1 oz. cream cheese	106	1 oz. Neufchatel cheese	70
1 cup vanilla ice cream	257	one-half cup vanilla ice cream	129
1 cup whole milk	170	skim milk	80
3 oz. dry salami	384	canned chicken	153
1 cup sour cream	454	low-fat yogurt	123
one-half cup frozen sweetened strawberries	139	frozen unsweetened strawberries	118
one-half cup oil-pack tuna	158	water-pack tuna	126
one-half cup syrup-pack canned fruit	90	drained/rinsed canned fruit	50
2 tablespoons mayonnaise	120	2 tablespoons mustard	24
4 tablespoons regular salad dressing	300	4 tablespoons diet salad dressing	80
1 piece fruit pie	410	1 piece fresh fruit	80

Notes:
REMEMBER: A calorie saved is a calorie burned...and 3,500 calories is equal to a pound of body fat.

Table C–2
Comparison of Regular and TRIMM* meals

Regular Food Item	Calories	TRIMM Food Item	Calories
Cream of tomato soup (8 oz.)	173	Tomato bouillon (8 oz.)	36
Meat loaf (4 oz.)	227	Meat loaf (3 oz.)	170
Brown Gravy (3 oz.)	130		
O'Brien potatoes	236	Parsley potatoes (1 med. 4½ oz.) NO BUTTER	99
Club spinach (w/eggs) ½ cup	153	Spinach (w/lemon) ½ cup	29
Tossed green salad (1 cup)	10	Tossed green salad (1 cup)	10
1000 Island Dressing (2 tbsp. or 1 oz.)	165	Diet 1000 Island Dressing (2 tbsp. or 1 oz.)	54
Yellow cake, chocolate frosting (3 inch squares.)	365	Fresh fruit (1 piece)	80
Milk, whole (8 fl. oz.)	159	Milk, skim (8 fl. oz.)	80
Totals	1618		558

Notes:
* TRIMM=The Right Image of Me in the Military.
The TRIMM Sample Meal is lower in fat and sugar—thus more "nutrient dense" and lower in calories

Table C–3
Sample 1200 Calorie Menu

	Average Calories
Breakfast	
1 cup fruit or juice (unsweetened)	100
1 ounce meat, cheese or egg (prepared without added fat)	80
1 slice toast (white, whole wheat, rye, or ¾ cup dry cereal)	70
1 cup skim milk	80
Total	330
Lunch	
2-ounce serving lean meat (or 1 oz. meat and 1 oz. cheese)	160
2 slices of bread or 1 bun or roll	140
1 serving cooked vegetables (leaves and stems—25 calories per ½ cup; starchy vegetables corn, lima beans—70 calories per one-third cup)	25
1 serving tossed vegetable salad	10
Vinegar or lemon juice	—
1 piece of *small* fresh fruit (or ¼ cup drained canned fruit)	50
Total	385
Supper	
3-ounce serving lean meat (no gravy)	225
one-half cup starch (potato, rice, pasta, dry beans or starchy vegetables)	80
1 serving cooked vegetables	25
1 serving tossed vegetable salad	10
Vinegar or lemon juice	—
1 piece of *small* fresh fruit (or ¼ cup drained canned fruit)	50
1 cup skim milk	80
Total	470
Choose Zero Calorie Beverages:	
Coffee or tea without cream or sugar, club soda, mineral water, iced water with lemon or lime.	0
Approximate Total for Day	1185

Table C–4
Sample 1500—Calorie Menu

	Average Calories
Breakfast	
1 cup fruit or juice (unsweetened)	100
1 ounce meat, cheese or egg, prepared without fat	80
1 slice toast (white, whole wheat, rye)	70
One-half cup cooked cereal or ¾ cup dry cereal (unsweetened)	70
1 cup skim milk	80
1 teaspoon margarine	45
Total	445
Lunch	
3 oz. serving lean meat (or 2 oz. meat and 1 slice cheese)	210
2 slices of bread or 1 bun or 1 roll	140
1 serving cooked vegetables (Leaves and stems 25 calories per ½ cup; starchy vegetables 70 calories per one-third cup)	25
1 tossed vegetable salad	10
1 tablespoon low-calorie dressing	30
1 piece of fresh fruit (medium to large)	80
1 small or ¼ cup drained fruit, 40 calories	
Total	520
Dinner	
3 oz. serving lean meat	225
one-half cup starchy vegetable (with small amount added fat)	100
Serving cooked vegetable	25
1 tossed vegetable salad	10
1 tablespoon low-calorie dressing	30
1 piece fresh fruit (medium to large)	80
1 cup skim milk	80
Total	550
Choose Zero Calorie Beverages:	
Coffee or tea without cream or sugar, club soda, mineral water, ice water with lemon or lime.	0
Approximate Total for Day	1515

Glossary

Section I
Abbreviations

AD
active duty

ADS
active duty support

ADT
active duty for training

AGR
Active Guard/Reserve

APFT
Army Physical Fitness Test

ARNG
Army National Guard

ARPERCEN
US Army Reserve Personnel Center

CAR
Chief, Army Reserve

CG
commanding general

CNGB
Chief, National Guard Bureau

DA
Department of the Army

DCSPER
Deputy Chief of Staff for Personnel

DCSLOG
Deputy Chief of Staff for Logistics

DOD
Department of Defense

FORSCOM
US Army Forces Command

HQDA
Headquarters, Department of the Army

HSC
US Army Health Services Command

IRR
Individual Ready Reserve

MACOM
major Army command

RC
Reserve Component

SADT
Special Active Duty for Training

TPU
troop program unit

TRADOC
US Army Training and Doctrine Command

TSG
The Surgeon General

USAR
US Army Reserve

Section II
Terms

Body composition
The human body is composed of two major elements: lean body mass (which includes muscle, bone, and essential organ tissue) and body fat. Body fat is expressed as a percentage of total body weight that is fat. For example, an individual who weighs 200 pounds and 18 percent body fat has 36 pounds of fat. Women generally have a higher percentage of body fat than men because of genetic and hormonal differences; thus, body fat standards differ among men and women by age groups.

Health care personnel
Trained physicians (military or civilian employees or contract personnel), physician's assistants, Registered Nurses, dieticians, and physical/occupational therapists under supervision of the unit surgeon or the commander of the medical treatment facility. For the purpose of this regulation, this term includes personnel of US forces and host nations.

Overweight
A soldier is considered overweight when his or her percent body fat exceeds the standard specified in paragraph 20*c* of this regulation.

Professional military or civilian schooling
This includes all individual training courses beyond Initial Entry Training (IET). It does not include unit training involving crews and teams. IET includes basic branch course or equivalent for officers; warrant officer entry course for non-prior service personnel; and basic training, AIT, OSUT, and OST for enlisted personnel.

Satisfactory progress
Progressing toward a point to meet the body fat standards described in paragraph 19*c*. Weight loss of 3–8 pounds per month is required for satisfactory progress.

Weight control
An individual program by which each soldier attains and maintains an acceptable weight and body composition through self-motivation or involvement in an official weight control program.

Section III
Special Abbreviations and Terms
There are no special terms.

Made in the USA
Monee, IL
05 November 2026

67808324R00085